CLOUD COMPUTING AND BIG DATA TECHNOLOGY

Second Edition

CLOUD COMPUTING
AND
BIG DATA TECHNOLOGY

Second Edition

Junxiu An

Chengdu University of Information Technology, China

Sian Jin

Temple University, USA

Lilang Wan

Chengdu University of Information Technology, China

World Scientific

NEW JERSEY · LONDON · SINGAPORE · BEIJING · SHANGHAI · TAIPEI · CHENNAI

Published by

World Scientific Publishing Co. Pte. Ltd.

5 Toh Tuck Link, Singapore 596224

USA office: 27 Warren Street, Suite 401-402, Hackensack, NJ 07601

UK office: 57 Shelton Street, Covent Garden, London WC2H 9HE

Library of Congress Cataloging-in-Publication Data

Names: An, Junxiu author | Jin, Sian author | Wang, Lilang author.

Title: Cloud computing and big data technology / authors, Junxiu An, Chengdu University of Information Technology, China, Sian Jin, Temple University, USA, Lilang Wang, Chengdu University of Information Technology, China.

Description: 2nd edition. | Singapore ; Hackensack, NJ ; London : World Scientific, 2025. | Includes index.

Identifiers: LCCN 2025010540 | ISBN 9789819812912 hardcover | ISBN 9789819812929 ebook for institutions | ISBN 9789819812936 ebook for individuals

Subjects: LCSH: Cloud computing | Big data

Classification: LCC QA76.585 .A53 2025 | DDC 005.7--dc23/eng/20250319

LC record available at https://lccn.loc.gov/2025010540

British Library Cataloguing-in-Publication Data

A catalogue record for this book is available from the British Library.

云计算与大数据技术应用第2版
Originally published in Chinese by China Machine Press
Copyright © China Machine Press 2022

For any available supplementary material, please visit
https://www.worldscientific.com/worldscibooks/10.1142/14306#t=suppl

Desk Editors: Aanand Jayaraman/Steven Patt

Typeset by Stallion Press
Email: enquiries@stallionpress.com

Preface

With the rapid development and widespread application of computers and the Internet, human society continues to advance toward digitization and networking. Cloud computing and big data technologies have emerged in response to these trends. The roots of cloud computing can be traced back to 1956, when Christopher Strachey published a paper on virtualization, formally introducing the concept of virtualization. Virtualization, the core of today's cloud computing infrastructure, laid the foundation for the evolution of cloud computing. The large-scale adoption of cloud computing has driven digital transformation across industries, thereby catalyzing the rise of big data.

This book is an updated edition of *Cloud Computing and Big Data Applications*, published in 2019. It primarily targets beginners interested in cloud computing and big data technologies, professionals engaged in application service software development, and students specializing in cloud computing and big data at vocational colleges and universities. Compared to the first edition, this revision features significant additions, deletions, and structural adjustments. For instance, content on data center deployment and site selection has been streamlined to enhance conciseness. A new Chapter 7 has been added, focusing on Docker container technology and the container orchestration platform Kubernetes, both widely adopted in enterprise IT today. Chapters 8–10 reorganize the introduction to big data frameworks, starting with batch processing to guide readers into the world of big data processing. Technical frameworks have also been updated — for example, Hadoop 3.0 with its architectural changes has been introduced, and Flink has been added to the Storm chapter for comparative learning — to present readers with more advanced technologies.

The book consists of 11 chapters:

Chapter 1: Overview of Cloud Computing introduces the definition, development background, infrastructure, service models, deployment modes, typical cloud computing products, recent advancements, and the current state of China's cloud computing industry.

Chapter 2: Overview of Big Data Technology covers the emergence of big data, the "4V" characteristics of big data, major applications, driving forces in the industry, key technologies, and typical big data computing architectures.

Chapter 3: Virtualization Technology provides a detailed introduction to virtualization, including its principles, common solutions, and practical applications.

Chapter 4: Data Centers and Cloud Storage discusses the basic concepts of data centers, trends in the era of cloud computing and big data, and focuses on cloud storage—its definition, architecture, foundational technologies, and features.

Chapter 5: Parallel Computing and Cluster Technology explores parallel computing fundamentals, clusters as the backbone of cloud infrastructure, classifications of parallel computing, related technologies, and hands-on MPI programming.

Chapter 6: OpenStack introduces the architecture and key modules of OpenStack, a powerful IaaS platform.

Chapter 7: Docker and Kubernetes covers Docker commands, image management, and container operations and provides an introduction to Kubernetes for container orchestration.

Chapter 8: Hadoop details the Hadoop ecosystem, including HDFS, MapReduce, HBase, and environment setup for distributed big data development.

Chapter 9: Storm and Flink explains Storm's principles, architecture, Storm–Yarn integration, environment configuration, and real-time stream processing applications, with a comparative introduction to Flink.

Chapter 10: Spark focuses on the Spark in-memory computing framework, covering its mechanisms, execution modes, RDDs, processing paradigms, and ecosystem.

Chapter 11: Cloud Computing Simulation introduces CloudSim, its modeling scenarios, and practical applications for simulating cloud environments.

This book is co-authored by Professor Junxiu An of Chengdu University of Information Technology and Assistant Professor Sian Jin of Temple University, as well as the following contributors from Chengdu University of Information Technology: Lilang Wan, Yimin Pan, Mingkun Yuan, and Liu Yuan. Specific authorship is as follows:

- Chapters 3, 6, 7, and 11: Lilang Wan and Junxiu An.
- Chapters 1, 8, 9, and 10: Yimin Pan and Sian Jin.
- Chapters 2, 4, and 5: Mingkun Yuan and Junxiu An.

Liu Yuan contributed to the review process.

The publication of this book is supported by the National Social Science Fund of China (Grant No. 21BSH016) and represents a milestone achievement of the Sichuan Provincial High-Level Social Science Research Team.

In addition to electronic courseware, source code, syllabi, exercises, and answers, this book collaborates with EduCoder (http://www.educoder. net) to provide online learning and lab resources. Scan the QR code provided at the end to access these resources, which include chapter-aligned experiments:

- Experiments 3, 6, and 7: Lilang Wan and Junxiu An.
- Experiments 1, 8, 9, and 10: Yimin Pan and Sian Jin.
- Experiments 2, 4, and 5: Mingkun Yuan and Junxiu An.

EduCoder, a massive open online practice (MOOP) platform, offers practical environments for experimental teaching, blended learning, first-class course development, and industry–academia collaboration. It covers programming, system design, electronics, cloud computing, big data, AI, blockchain, and more, serving as a national platform for top-tier course development.

Despite our rigorous efforts to ensure accuracy, the ever-evolving nature of technology and the limits of our expertise may have led to oversights. We warmly welcome feedback and suggestions via email at 86631589@qq.com.

Online Lab Resources for This Book	**EduCoder WeChat Official Account**

Junxiu An
Chengdu University of Information Technology
March 2025

About the Authors

An Junxiu is a professor at Chengdu University of Information Technology, as well as a senior visiting scholar and master's supervisor. She serves as the academic leader of the Sichuan Provincial Key Laboratory of Software Automatic Generation and Intelligent Service (Domain Ontology and Big Data) and is the head of the Parallel Computing and Big Data Research Institute. With extensive experience in research and teaching in the fields of data science and big data technology, she has published over 50 papers in related areas and authored nearly 20 monographs or textbooks on cloud computing, big data, and artificial intelligence. Additionally, she is an evaluation expert for the National Natural Science Foundation of China, the Sichuan Science and Technology Project, and the Chengdu Science and Technology Project.

Sian Jin is an assistant professor at Temple University's Department of Computer and Information Sciences. He received his PhD in computer engineering from Indiana University in 2023. He received his bachelor's degree in physics from Beijing Normal University in 2018. His research interests include high-performance computing (HPC), data reduction, and lossy compression to improve performance in scientific data analytics and management, as well as in large-scale machine learning and deep learning. In the past five years, he has published over 20 papers in top conferences and journals, including SC, PPoPP, VLDB, ICDE, EuroSys, HPDC, and ICS.

Wan Lilang is a master's student at Chengdu University of Information Technology, focusing on non-stationarity and distribution shifts in time series data. He has contributed to the initial editions of four textbooks published by national-level publishing houses and has published four papers in related fields.

Contents

Chapter 1

Overview of Cloud Computing

Cloud computing is a mode based on the increase, access, and delivery of Internet-related services, which usually involves the provision of dynamic, scalable, and often virtualized resources over the Internet. It is the fusion and development of parallel computing, distributed computing, and grid computing, as well as the result of commercial implementations of the hybrid evolution of concepts such as virtualization, utility computing, and service-oriented architecture (SOA). This chapter introduces cloud computing, its development background and characteristics, its infrastructure, its business models, and its main service models.

1.1. What is Cloud Computing?

With the continuous development of computer technology, cloud computing has become a new force driving the transformation of social productivity. So, what is cloud computing? What are its characteristics? The following discusses these questions.

1.1.1. *Definition of cloud computing*

Currently, there is no unified definition of cloud computing in the industry. Leading players in the cloud computing industry such as Google, Microsoft, and other IT manufacturers and research institutions have provided the following definitions and understandings of cloud computing based on their own interests and research perspectives:

(1) *Wikipedia*: Cloud computing is a dynamically scalable computing model that delivers virtualized resources as services to users over computer networks. Cloud computing typically includes Infrastructure as a Service (IaaS), Platform as a Service (PaaS), and Software as a Service (SaaS).
(2) *Google*: Cloud computing is to place all computing and applications in the "cloud", and terminal devices do not need to install any software to share programs and services via the Internet.
(3) *Microsoft*: Cloud computing is the computation of "cloud + terminal", where computing resources are dispersed and distributed. Some resources are placed in the cloud, some on user terminals, and some with partners. Ultimately, users choose the appropriate computing resources.
(4) *International Data Corporation* (*IDC*): Cloud computing is a novel IT development, deployment, and delivery model that can provide products, services, and solutions in real time over the Internet.
(5) *National Institute of Standards and Technology* (*NIST*): Cloud computing is a ubiquitous, convenient, Internet-accessible, customizable pool of IT resources (including networks, servers, storage, applications, and services) shared on a pay-per-use basis. It enables the rapid provisioning and release of computing resources with minimal management or interaction with service providers. And this is the widely accepted definition of cloud computing at present.
(6) The United States Federal Cloud Computing Strategy Report defines four types of clouds:

- *Public Cloud*: It provides cloud computing services to the general public and public groups, such as the Amazon Cloud platform and Google App Engine (GAE). The public cloud has many advantages, but the biggest drawback is the difficulty in ensuring data privacy.
- *Private Cloud*: It provides cloud computing services for industries/organizations internally, such as cloud platforms used internally by government agencies, enterprises, and schools. Private clouds can better address data privacy issues. For industries or organizations with particularly high data privacy requirements,

building a private cloud is an inevitable choice. For example, Windows Azure is a private cloud platform management and service software.

- *Community Cloud*: It provides cloud computing platforms for users within community organizations. For example, the Nebula cloud platform of the National Aeronautics and Space Administration (NASA) provides rapid IT access services for NASA researchers.
- *Hybrid Cloud*: A hybrid cloud platform that includes more than two of the above cloud computing types.

(7) *China's Definition of Cloud Computing*: In March 2012, in the Government Work Report of the State Council, cloud computing was designated as a strategic emerging industry for the country, and a definition was provided: Cloud computing is a mode based on the increase, access, and delivery of Internet-related services, which usually involves the provision of dynamic, scalable, and often virtualized resources over the Internet. Cloud computing is the product of the development and convergence of traditional computer and network technologies, which means that computational capabilities can also be circulated as a commodity through the Internet.

Cloud computing can be classified into two categories: broad and narrow. In the narrow sense, cloud computing refers to the delivery and usage model of IT infrastructure, that is, resources are obtained as needed and expanded over the network. In the broad sense, cloud computing refers to the usage and delivery model of services, that is, corresponding services are obtained as needed and expanded over the network.

In short, cloud computing is a computing model that provides dynamically scalable virtualized resources as services over the Internet. The resources in cloud computing are distributed and dynamically scalable through virtualization technology. Cloud computing is a service provided over the Internet with a Service-Level Agreement (SLA). The SLA is a commercial guarantee contract between cloud service providers and customers, rather than a general service promise. End users do not need to understand the details of the infrastructure in the "cloud", nor do they need corresponding expertise or direct control. They only need to focus on what resources they truly need and how to obtain corresponding services over the network.

1.1.2. *Conceptual model of cloud computing*

According to the concept of cloud computing, it is evident that the essence of cloud computing is the application on the network. The conceptual model of its business implementation is illustrated in Figure 1.1. Cloud computing encompasses multiple layers of significance, as follows:

(1) *Public Nature of Users*: Cloud computing caters to various users, including enterprises, government departments, academic institutions, individuals, as well as "users" such as application software and middleware platforms. Middleware refers to an independent system software or service program that distributed application software uses to share resources between different technologies. Positioned above the operating system of client/server systems, middleware manages computer resources and network communication, serving as the software that connects two independent applications or systems together.

(2) *Diversity of Devices*: The devices used to provide services in cloud computing are diverse, including various scales of servers, hosts, and storage devices, as well as various types of terminal devices such as computers, smartphones, various smart sensors, and Radio Frequency

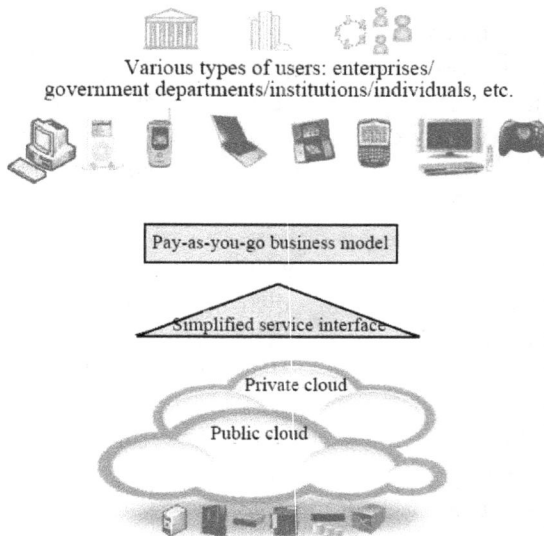

Fig. 1.1.   Conceptual model of cloud computing.

Identification (RFID) devices. RFID is a communication technology that identifies specific targets and reads/writes relevant data through radio signals, eliminating the need to establish mechanical or optical contact between the identification system and the specific target.

(3) *Service-Oriented Business Model*: The service characteristics of cloud computing are reflected in two aspects: simplified and standardized service interfaces, and a pay-as-you-go business model.

(4) *Flexibility of Delivery*: Cloud computing can be utilized as a public facility to provide societal services, known as the "public cloud", or it can be employed as a centralized computing platform for enterprise informatization, known as the "private cloud".

1.1.3. *Characteristics of cloud computing*

Cloud computing has the following characteristics compared to traditional computer systems:

(1) *Massive Parallel Computing Capability*: Based on the powerful and cost-effective computing capabilities on the cloud, cloud computing offers computing services for large-grained applications that traditional computing systems or user terminals cannot achieve. The computing resources of cloud computing systems include CPU processing resources, storage resources, and network bandwidth. Typically, private clouds in enterprises have hundreds to thousands of servers, and some even have millions of servers.

(2) *Resource Virtualization and Elastic Scheduling*: The resource pool of cloud computing systems includes storage, processors, memory, network bandwidth, and other resources. They are allocated on demand to provide computing resources for small-grained applications and realize resource sharing. Moreover, the scale of cloud computing systems can dynamically scale up or down to meet different applications and user demands. At the same time, different physical machine and virtual machine resources in cloud computing systems can be dynamically allocated according to customer requirements. The resources obtained by customers may come from cloud computing resources in Beijing or Shanghai. Virtualization technology is also one of the core technologies of cloud computing, which includes network virtualization, storage virtualization, server virtualization, operating system virtualization, and application virtualization.

(3) *Enormous and Rapidly Growing Data Volume*: In the cloud computing environment, people are both users and creators of information, leading to a dramatic increase in the volume of information on the Internet. Consequently, how to leverage these data to provide better services has become a current research focus, resulting in typical big data processing technologies such as Hadoop, Spark, and Storm. These technologies are closely intertwined with cloud computing.

Cloud computing also has several other characteristics: Firstly, high reliability, which means that cloud computing employs measures such as data replication for fault tolerance and interchangeable homogeneous computing nodes to ensure high reliability; secondly, universality, meaning that the same cloud can support the operation of different applications; and lastly, cost-effectiveness, which refers to the relatively low prices of cloud computing, making it more cost-effective for users.

1.2. Background of Cloud Computing Technology Development

From a technological perspective, the emergence of cloud computing is attributed to the rapid growth of network bandwidth and the swift development of the Internet. The rapid expansion of network bandwidth has made it possible to process data quickly in the Internet environment. The Internet provides a new way of browsing and exchanging data, where users are not only consumers but also creators of information. Massive amounts of data are generated by users. The Internet has become an indispensable infrastructure in our daily lives and has been rapidly developed. The exponential increase in smart terminal devices and Internet users has driven the emergence of cloud computing, which has subsequently evolved into an essential infrastructure.

(1) *The Rapid Development of the Internet*: The first wave of informatization revolution, namely the computer revolution, occurred in the 1960s. Many traditional enterprises followed this wave of informatization and extensively applied computers to their businesses. The second wave of informatization revolution, known as the Internet revolution, took place in the 1990s. On September 14, 1987, China sent its first email: "Across the Great Wall we can reach every corner

```
DATE:      MON, 14 SEP 87 21:07 CHINA TIME
FROM:      "MAIL ADMINISTRATION FOR CHINA" <MAIL@ZE1>
TO:        ZORN@GERMANY, ROTERT@GERMANY, WACKER@GERMANY, FINKEN@UNIKA1
CC:        LHL@PARMESAN.WISC.EDU, FARBER@UDEL.EDU,

           JENNINGS%IRLEAN.BITNET@GERMANY, CIC%RELAY.CS.NET@GERMANY, WANG@ZE1,

           RZLI@ZE1
SUBJECT:   FIRST ELECTRONIC MAIL FROM CHINA TO GERMANY


"UEBER DIE GROSSE MAUER ERREICHEN WIE ALLE ECKEN DER WELT"

"ACROSS THE GREAT WALL WE CAN REACH EVERY CORNER IN THE WORLD"


DIES IST DIE ERSTE ELECTRONIC MAIL, DIE VON CHINA AUS UEBER RECHNERKOPPLUNG
IN DIE INTERNATIONALEN WISSENSCHAFTSNETZE GESCHICKT WIRD.

THIS IS THE FIRST ELECTRONIC MAIL SUPPOSED TO BE SENT FROM CHINA INTO THE
INTERNATIONAL SCIENTIFIC NETWORKS VIA COMPUTER INTERCONNECTION BETWEEN
BEIJING AND KARLSRUHE, WEST GERMANY (USING CSNET/PMDF BS2000 VERSION).

   UNIVERSITY OF KARLSRUHE          INSTITUTE FOR COMPUTER APPLICATION OF
-INFORMATIK RECHNERABTEILUNG-       STATE COMMISSION OF MACHINE INDUSTRY
         (IRA)                                     (ICA)

PROF. WERNER ZORN                   PROF. WANG YUN FENG
MICHAEL FINKEN                      DR. LI CHENG CHIUNG
STEFAN PAULISCH                     QIU LEI NAN
MICHAEL ROTERT                      RUAN REN CHENG
GERHARD WACKER                      WEI BAO XIAN
HANS LACKNER                        ZHU JIANG
                                    ZHAO LI HUA
```

Fig. 1.2.   China's first email.

in the world", marking the beginning of Chinese Internet usage, as shown in Figure 1.2. At that time, the communication speed of the Internet was 300 bit/s.

In November 1997, the China Internet Network Information Center (CNNIC) released the first "Statistical Report on Internet Development in China": China had a total of 299,000 Internet-connected computers, with 620,000 Internet users, 4,066 domain names registered under ".CN", approximately 1500 WWW sites, and an international outbound bandwidth of 25.408 Mbit/s. By January 1999, the international outbound bandwidth had increased to 143 Mbit/s. As of December 2020, China's international outbound bandwidth had reached 11,511,397 Mbit/s, showcasing the rapid pace of network development. In 2003, the rise of mobile Internet began,

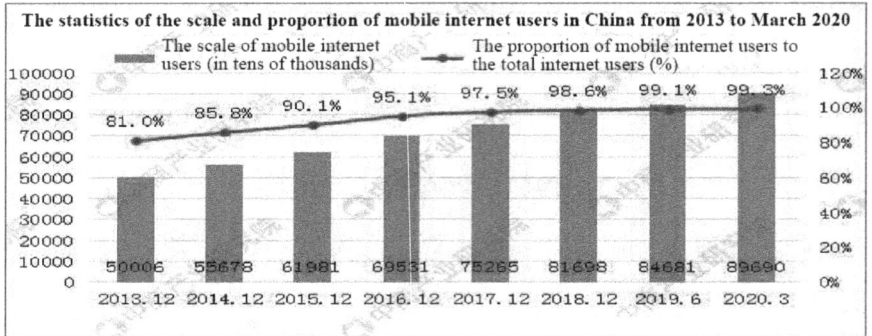

Fig. 1.3.   The scale and proportion of mobile internet users.

and with the emergence of 3G and 4G, the speed of mobile Internet achieved rapid improvement. If 3G and 4G fully connected people, then 5G can achieve the Internet of Everything, especially providing a solid foundation for the realization of virtual reality, autonomous driving, and telemedicine. According to a report from the CNNIC, as of December 2020, China's Internet users reached 989 million, with a penetration rate of 70.4%. Mobile Internet users accounted for 99.3%, and the habit of mobile payment had been formed. Figure 1.3 shows the scale and proportion of mobile Internet users.

The core of cloud computing lies in network-based applications, where the bandwidth of network access directly determines the quality of cloud computing platforms used by enterprises. While cloud computing services are simple, lightweight, and convenient to use, they consume significant amounts of network bandwidth behind the scenes, leading to a substantial increase in Internet traffic. Cloud computing is a globally accessible resource structure, implying that users adopting cloud computing architectures will be across a wide area network.

(2) *The Invention and Development of the World Wide Web*: In 1989, Tim Berners-Lee, working at the European Organization for Nuclear Research (CERN), invented the World Wide Web (WWW) out of the necessity for high-energy physics research, as depicted in Figure 1.4. Four years later, the American company Netscape Communications Corporation launched the Mosaic web browser, which instantly

Fig. 1.4 .   The inventor of the WWW and the world's first website.

became popular worldwide. The birth of the WWW has brought revolutionary changes to the exchange and dissemination of global information, opening the convenient gateway for people to access information through the Internet.

The WWW is not synonymous with the Internet; rather, it is just one of the services that the Internet can provide, operating as a service reliant on the Internet. Perhaps surprisingly, despite the countless wealth created by Tim Berners-Lee's invention of the WWW over more than 20 years, the creator himself insisted on not patenting the World Wide Web and on making it freely available to everyone. Due to his outstanding contribution, he is hailed as the "Father of the Internet".

The Web 1.0 era began in 1994, characterized by the widespread use of static HTML web pages to publish information and the use of browsers to access information. At this time, information transmission was primarily one-way. The essence of Web 1.0 is aggregation, association, and search, with the aggregation of vast, unordered Internet information. Web 1.0 only addressed the need for people to search for aggregate information, without addressing the need for communication, interaction, and participation between individuals. During this period, notable Internet companies such as Baidu, Google, and Amazon were born.

At the beginning of the 21st century, just as the dot-com bubble burst, the rise of Web 2.0 (which began in March 2004) ushered in a new peak of development for the Internet. In the era of Web 2.0, software is treated as a service, and the WWW has evolved into a mature platform that provides network applications for end users. It emphasizes user participation, online collaboration, networked data storage, social networking, Really Simple Syndication (RSS) applications, and file sharing. At this time, information transmission on the Internet became bidirectional, with users being both consumers and creators of information. The Web 2.0 model greatly stimulates creativity and innovation, making the Internet vibrant.

In the Web 2.0 era, social networking sites such as Flickr, Myspace, Facebook, YouTube, blogs, and wikis have far surpassed traditional portal websites in terms of traffic. The high number of users and their high degree of engagement are characteristics of these websites. Therefore, how to effectively serve such a large user base and provide them with convenient and efficient services during participation becomes a problem that these websites must address. To address the challenges of large-scale website traffic, high concurrency, and massive data, enterprises generally consider business decomposition and distributed deployment. They may isolate less related businesses and deploy them on different machines, thereby achieving large-scale distributed systems. This has also driven the emergence of cloud computing and big data.

(3) *Evolution of the Information Industry*: Throughout the entire history of information technology development (as illustrated in Figure 1.5), two important core driving forces have played key roles in the development of the information industry at different times: hardware driving force and network driving force. The comparison and changes of these two driving forces determine the periods when different products appear in the industry and the times when different forms of enterprises emerge and disappear. The era driven by hardware gave birth to enterprises such as IBM, Microsoft, and Intel, while the era driven by networks gave birth to enterprises such as Google, Yahoo, and Amazon.

With the onset of the third wave of the information revolution in 2010, known as the mobile Internet revolution, the world officially entered the era of big data. Concurrently, the Google server cluster, built on the foundation of the Google File System (GFS), provided Google with powerful search speed and processing capabilities. How to effectively utilize these server resources to offer robust computing power and a variety of services to more enterprises or individuals

Web 2.0

Network — the Core Driving Force

BAT

Google, Yahoo, Amazon

Hardware — the Core Driving Force

Microsoft: Intel

IBM

Centralized computing

Mainframe

Minicomputer

PC

UNIX Workstation/Server

Local Area Network

Client/Server (C/S)

Browser/Server (B/S)

Utility Computing

Grid Computing

Internet of Things (IoT)

SaaS

PaaS

IaaS

Birth of Computers   Personal Computer   Internet   Cloud Computing

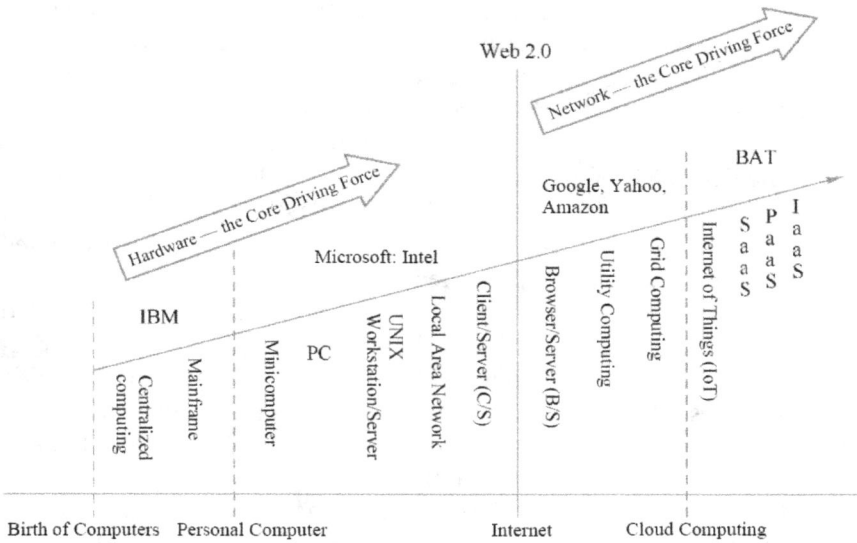

Fig. 1.5.   Evolutionary roadmap of information industry development.

became a critical consideration for companies with massive server resources like Google. Thus, cloud computing came into being.

(4) *The Emergence of Cloud Computing*: The computational capability provided by computers is dependent on hardware resources. When computational demands exceed capacity, it becomes necessary to continuously augment hardware resources. Conversely, during periods of low demand and ample computational capacity, hardware resources remain underutilized. Failure to share and utilize these resources represents a significant waste. Cloud computing emerged as a novel computing service model aimed at addressing this issue. Its fundamental approach is to centralize computational resources to deliver vast computational power while ensuring convenience and flexibility in usage.

In the IT field, any new concept requires both business and technological impetus. With the continuous development and maturation of distributed storage, parallel computing, virtualization technology, and Internet technology, it is possible to provide services based on the Internet. In 1983, Sun Microsystems introduced the concept of "cloud computing", stating that "The Network is the Computer". However, due to the lack of a specific infrastructure proposal, this concept did not gain widespread recognition. On August 9, 2006, Eric Schmidt, then CEO of Google, first proposed the concept and architectural framework of cloud computing at the Search Engine Strategies conference (SES San Jose 2006). He integrated Google's distributed file storage system, GFS, presented at the 2003 Symposium on Operating Systems Principles (SOSP), MapReduce distributed processing technology presented at the 2004 Operating Systems Design and Implementation (OSDI), and the BigTable distributed data storage system introduced the same year to propose the infrastructure of cloud computing. GFS addressed the challenge of storing large files. MapReduce tackled parallel computing issues. And BigTable resolved the storage problem of massive non-relational databases. These three papers only provided conceptual ideas and did not open-source the code. The concept of MapReduce was originally proposed by the "Father of Artificial Intelligence", John McCarthy, in 1960, when he predicted that "[i]n the future, computers will be provided to the public as a utility". Therefore, the initial concept of cloud computing can be traced back to an earlier period. As shown in Figure 1.6, two representative figures proposed the concept of cloud computing. In 2008, cloud computing entered China, and in 2009, China's first Cloud Computing Conference was held. In 2012, China's government work report provided a definition of cloud computing.

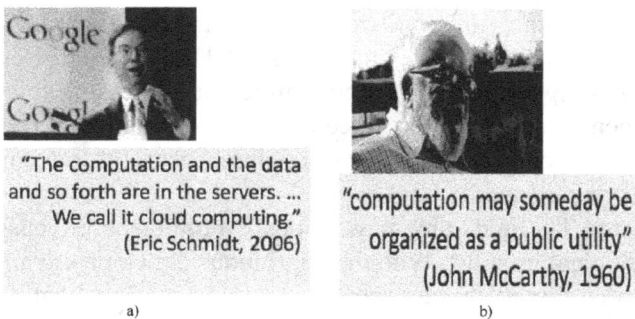

Fig. 1.6.   Representatives who proposed cloud computing: (a) Eric Schmidt and (b) John McCarthy, the father of artificial intelligence.

1.3. Typical Cloud Computing Infrastructure

After understanding the concept and development background of cloud computing technology, let's take Google's cloud computing architecture as an example to introduce a typical cloud computing infrastructure.

Google's cloud computing technology is actually developed for its most important search applications. In response to the massive scale of internal network data, Google has proposed a complete set of distributed parallel cluster infrastructure. Its data centers use inexpensive Linux PCs to form clusters. Software is utilized to handle the frequent node failures in the cluster, thus forming Google's cloud computing infrastructure.

Google's cloud computing infrastructure consists of three interrelated yet tightly integrated systems: GFS, MapReduce programming model tailored for Google applications, and the large-scale distributed database BigTable, as shown in Figure 1.7.

(1) *GFS is a distributed file system built on clusters*. To meet its rapidly growing data processing needs, Google has optimized the file system, which has addressed issues such as accessing huge files, a significantly higher read-to-write ratio, and node failures in the cluster. GFS defaults to dividing large files into 64 MB blocks, which are distributed across machines in the cluster and stored using the Linux file system. Additionally, each file block has at least three redundant copies to solve this problem.

(2) *MapReduce is a distributed parallel programming model*. Google devised the MapReduce programming model to simplify programming in distributed systems. Users only need to provide their own Map

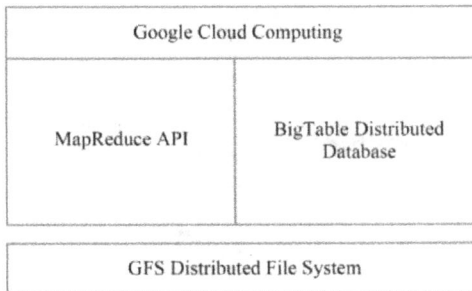

Google Cloud Computing	
MapReduce API	BigTable Distributed Database
GFS Distributed File System	

Fig. 1.7. Google's cloud computing architecture.

and Reduce functions, enabling large-scale distributed parallel data processing on the cluster. The Map function breaks down the input into intermediate Key/Value pairs, while the Reduce function aggregates these pairs into the final output. These functions are provided by developers and executed across the cluster, with results stored in GFS.

(3) *BigTable is a distributed, large-scale database management system.* Due to the vast amount of semi-structured data that Google applications need to handle, Google constructed BigTable, a large-scale database system with weak consistency requirements. It features a sparse, distributed, persistent, multidimensional-sorted data model stored in the form of key/value pairs. BigTable is not a relational database; as its name suggests, it is essentially a massive table designed for storing semi-structured data.

The above are the three main components of Google's internal cloud computing architecture. In addition to these three components, Google has also built other cloud computing components, including a domain-specific language, a distributed program scheduler, and a distributed lock service known as the Chubby mechanism.

Figure 1.8 illustrates the core technologies of Google Cloud Computing. Among them, data processing adopts the MapReduce parallel programming model; large file storage adopts GFS; large-scale database management systems adopt BigTable; and cloud computing services adopt GAE. The widely used Hadoop is an open-source implementation of Google's core technologies, such as MapReduce, GFS, and BigTable, supported by the Apache Software Foundation.

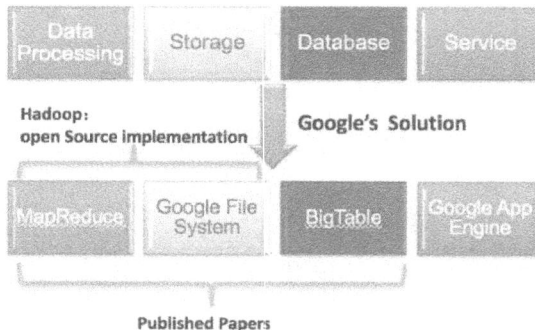

Fig. 1.8.   Google's core cloud computing technologies.

1.4. Main Service Models of Cloud Computing

From the perspective of user experience, cloud computing is primarily divided into three service models: IaaS, PaaS, and SaaS. SaaS focuses on software services, providing software program services over the network; PaaS focuses on platform services, providing services through service platforms or development environments; and IaaS emphasizes hardware resource services, focusing on the sharing of computing resources, allowing consumers to access services from a comprehensive computing infrastructure via the Internet.

1.4.1. *IaaS*

Based on the IaaS model, virtual computing and data centers can consolidate computing units, memory, I/O devices, bandwidth, and other computer infrastructure into a virtual resource pool, providing services over the network. IaaS offers computing resources and infrastructure services that are close to bare metal (physical machines or virtual machines).

A typical representative of IaaS is Amazon Web Service (AWS), Amazon's cloud computing service, as illustrated in Figure 1.9. It

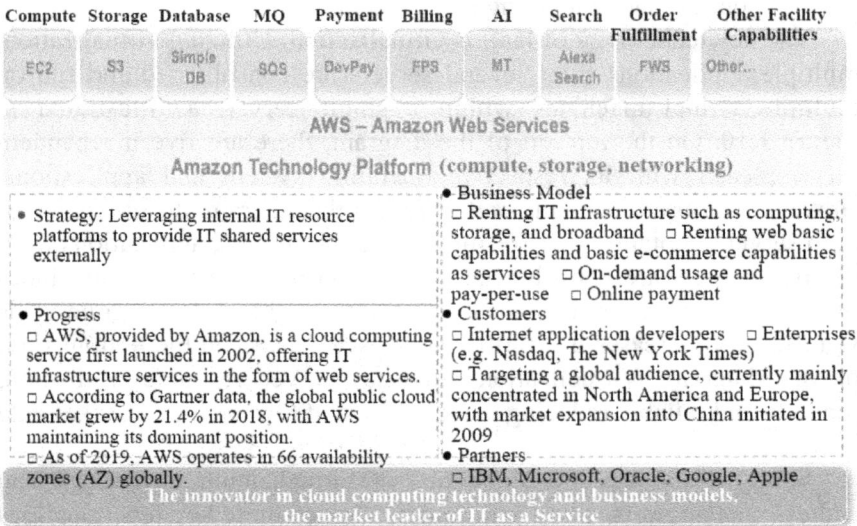

Compute	Storage	Database	MQ	Payment	Billing	AI	Search	Order Fulfillment	Other Facility Capabilities
EC2	S3	Simple DB	SQS	DevPay	FPS	MT	Alexa Search	FWS	Other...

AWS – Amazon Web Services

Amazon Technology Platform (compute, storage, networking)

- Strategy: Leveraging internal IT resource platforms to provide IT shared services externally

- Progress
  □ AWS, provided by Amazon, is a cloud computing service first launched in 2002, offering IT infrastructure services in the form of web services.
  □ According to Gartner data, the global public cloud market grew by 21.4% in 2018, with AWS maintaining its dominant position.
  □ As of 2019, AWS operates in 66 availability zones (AZ) globally.

- Business Model
  □ Renting IT infrastructure such as computing, storage, and broadband   □ Renting web basic capabilities and basic e-commerce capabilities as services   □ On-demand usage and pay-per-use   □ Online payment
- Customers
  □ Internet application developers   □ Enterprises (e.g. Nasdaq, The New York Times)
  □ Targeting a global audience, currently mainly concentrated in North America and Europe, with market expansion into China initiated in 2009
- Partners
  □ IBM, Microsoft, Oracle, Google, Apple

The innovator in cloud computing technology and business models, the market leader of IT as a Service

Fig. 1.9. Amazon's cloud service, AWS.

provides two typical cloud computing platforms: Elastic Computing Cloud (EC2) and Simple Storage Service (S3). EC2 performs computing functions, allowing users to deploy their own system software and develop and release application software on this platform. S3 handles storage and computing functions, where the basic unit in S3 is a bucket, serving as a container for storing files. Each bucket and file within it is assigned a URI address by S3, allowing users to access files via the HTTP or HTTPS protocol. The services charged include storage servers, bandwidth, CPU resources, and monthly rental fees. The monthly fee is similar to telephone monthly fees, where storage servers and bandwidth are charged based on capacity, and CPU usage is charged based on the duration (hours) of calculation volume.

As early as 2007, The New York Times rented the Amazon cloud computing platform to convert 11 million newspaper articles into PDF files from 1851 to 1922, which were made available for free online access to readers. It rented a total of 100 EC2 nodes, running for 24 hours and processing 4 TB of raw scanned newspaper images, generating 1.5 TB of PDF files. The cost per node per hour was 10 cents, resulting in a total cost of only $240 for the entire computing task (100 nodes × 24 hours × $0.10). If The New York Times had used its own servers, it would have taken several months and incurred expensive costs! Therefore, when users need to run batches of programs but lack suitable software and hardware environments, cloud computing is an excellent choice.

The key technology of IaaS is virtualization. Utilizing virtualization, multiple applications from several servers are consolidated and run on multiple virtual machines within a single server, as illustrated in Figure 1.10. On the top left of the diagram, there are five independent servers, each with its respective operating system and applications. However, as observed on the bottom left of the diagram, the utilization of each server is notably low. To fully utilize the servers, applications from the five servers are consolidated and run on multiple virtual machines within a single server, as depicted on the top right of the diagram. This significantly increases utilization rates, as indicated on the bottom right of the diagram. Virtualization enhances server resource utilization, securely and reliably reducing the Total Cost of Ownership (TCO) of data centers.

Some of the key functionalities of virtualization technology can address challenges faced by data centers. One of these key functionalities is partitioning. Partitioning refers to the virtualization layer's ability to

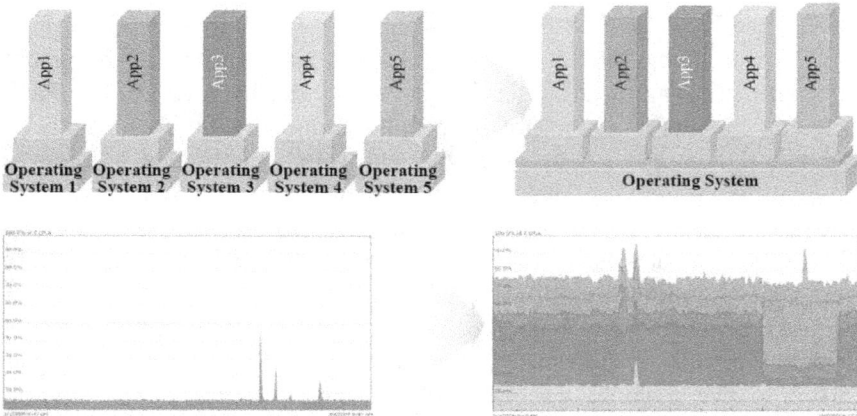

Fig. 1.10. One server virtualizes and consolidates five applications, thereby enhancing CPU utilization.

allocate server resources for multiple virtual machines. Each virtual machine can run a separate operating system (whether identical or different), enabling multiple applications to run on a single server. Each operating system only "sees" the "virtual hardware" provided by the virtualization layer (such as virtual network interface cards and SCSI cards), making it believe that it is running on its own dedicated server.

Another major functionality of virtualization is isolation. If a virtual machine crashes or experiences a failure (such as OS failure, application crashes, or driver failures), it does not affect other virtual machines on the same server. Viruses, worms, and other threats within one virtual machine are isolated from others, as if each virtual machine were on a separate physical machine. Virtualization technology also enables resource control to provide performance isolation, allowing each virtual machine to specify minimum and maximum resource usage to ensure that one virtual machine does not consume all resources, avoiding leaving none available for other virtual machines in the same system.

The third important functionality of virtualization is encapsulation. Encapsulation means to store the entire virtual machine (including hardware configuration, BIOS configuration, memory state, disk state, I/O device state, and CPU state) in a small set of files independent of physical hardware, making copying and moving virtual machines as simple as copying and moving files.

1.4.2. *PaaS*

PaaS is a business model that provides the runtime and development environment of application services as a service. In other words, PaaS offers developers an environment to build applications without the need to concern themselves with underlying hardware, enabling easy access to many necessary services during application development. For instance, if software developers want to upload a large file to the network and allow 35,000 users to access it over a period of two months, they can utilize Amazon's CloudFront platform to accomplish this.

GAE provides a PaaS cloud computing service platform tailored specifically for software developers. It consists of a Python application server cluster, BigTable database access, and GFS data storage services, offering developers an integrated, hosted, and automatically scalable online application service. Users write their applications, and Google provides all the platform resources needed for application execution and maintenance. The working principle of the cloud computing service platform, GAE, is illustrated in Figure 1.11. On the GAE platform, developers need not worry about the resources required for application execution, as GAE takes care of everything. This makes it easier for developers to create and upgrade online applications without having to expend effort on system management and maintenance.

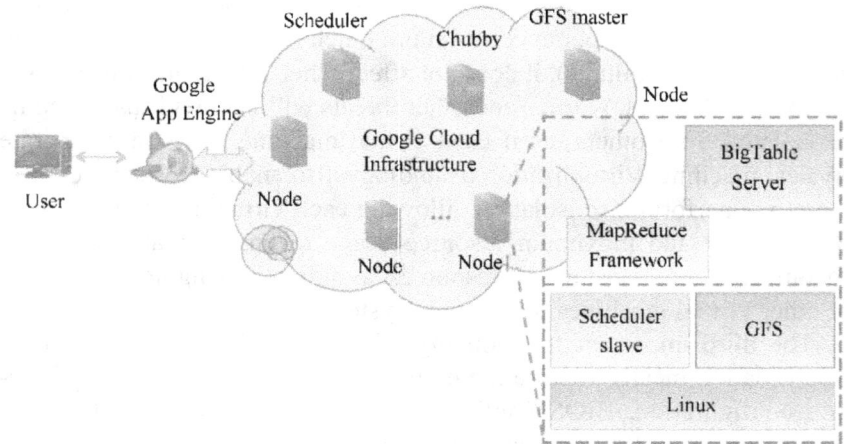

Fig. 1.11.   The operational principles of the cloud computing service platform, GAE.

The service provided by GAE allows developers to compile Python-based applications and host them using Google's infrastructure for free (with a maximum storage space of 500 MB). Beyond this limit, Google charges users based on CPU core usage time and storage space consumption according to certain criteria.

GAE differs from Amazon's S3, EC2, and SimpleDB in that the latter directly provide a range of hardware resources for users to choose from.

There are two key technologies in PaaS: distributed parallel computing and large-file distributed storage. Distributed parallel computing technology aims to fully utilize widely deployed ordinary computing resources to achieve large-scale computation and application goals. It transforms traditional computations into parallel computations, providing parallel services to customers. Large file distributed storage ensures the security and operability of massive data stored on clusters of inexpensive and untrusted nodes.

1.4.3. *SaaS*

SaaS is an application model that provides software services over the Internet, offering various application software services. Users only need to pay for usage time and scale, without the need to install corresponding application software. They can simply run the software (application service) by opening a browser, without requiring additional server hardware, thus achieving on-demand customization of software. From the user's perspective, SaaS eliminates expenses on server and software licensing. From the supplier's point of view, maintaining only one application reduces costs. SaaS mainly targets ordinary users.

SaaS is a software application model that emerged in the 21st century with the development of Internet technology and the maturity of application software. The SaaS service model differs significantly from the traditional sales of software perpetual licenses. SaaS adopts a software leasing model, which is considered the future trend in software management.

For the vast majority of small and medium-sized enterprises, SaaS is the best way to implement informatization using advanced technology. Enterprises can use information systems via the Internet without needing to purchase software or hardware, build data centers, or recruit IT personnel. Just like turning on a tap for water, enterprises can lease software services from SaaS providers according to their actual needs.

Technical Mode	■ Users can utilize on-demand customized software services and pay according to the amount and duration of customized software usage. ■ Users access the required services through a web browser without having to invest significant effort in building and maintaining IT infrastructure.
Hosting Platform	■ Directly deployed on underlying physical resources ■ Deployed on an IaaS (Infrastructure as a Service) platform ■ Deployed on a PaaS (Platform as a Service) platform
Development Technologies	■ Developing and deploying SaaS (Software as a Service) services on a PaaS platform ■ Developing under the SOA (Service-Oriented Architecture) framework

Significance

Transforming software provision into an internet-based service

Lowering the barriers for users to access software applications

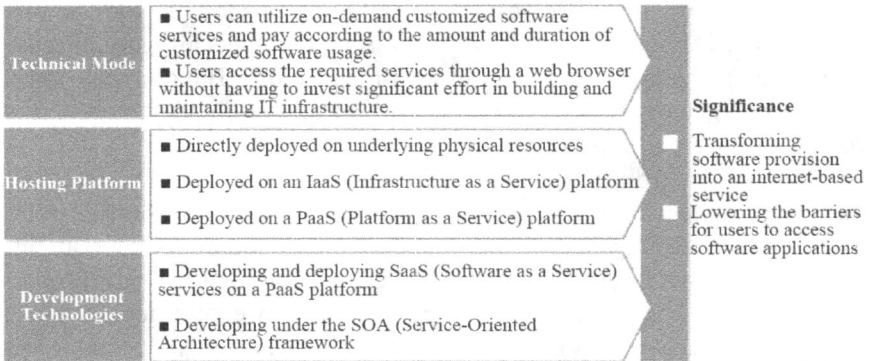

Fig. 1.12.   The key technology of SaaS: multi-tenancy technology approach.

Typical products of SaaS include Salesforce.com, Alibaba Software, MainOne, eAbax, Grow Force, Digital China, BizNavigator, YouShang.com, 800APP, and bibisoft.cn. Among them, Salesforce.com is the global leader in on-demand Customer Relationship Management (CRM) solutions, with Alibaba Software ranking second worldwide.

The key technology of SaaS is multi-tenancy. Cloud computing requires hardware and software resources to be shared more efficiently, with good scalability. Each user should be able to customize configurations according to their needs without affecting the usage of other users. Multi-tenancy technology is the key technology in cloud computing environments that meets the above requirements, as illustrated in Figure 1.12.

1.4.4. *The relationship among the three service models*

The previous sections have outlined three service models of cloud computing from the perspective of user experience: IaaS, PaaS, and SaaS. The relationship between them can be analyzed from two perspectives.

(1) *User Experience Analysis*: From the perspective of user experience, their relationship is independent because they cater to different types of users. SaaS primarily targets ordinary users who, at any time or place, can directly use applications running in the cloud through a browser, without the hassle of software installation. Moreover, they can avoid high initial investments in software and hardware. PaaS

mainly serves developers. To support the entire PaaS platform, suppliers need to provide four major functionalities: a user-friendly development environment, rich services, automatic resource scheduling, and precise management and monitoring. IaaS primarily serves system administrators with specialized knowledge. IaaS suppliers need to manage infrastructure in seven aspects to provide resources to users. These aspects include resource abstraction, resource monitoring, load management, data management, resource deployment, security management, and billing management.

(2) *Technical Analysis*: The service hierarchy of cloud computing is based on service types. From a technical perspective, there is a certain hierarchical relationship among them, namely, SaaS is built on PaaS, and PaaS is built on IaaS. However, this relationship is not simplistic. This is because SaaS can be built on top of PaaS or deployed directly on IaaS. PaaS can be constructed on top of IaaS or directly on physical resources. In other words, each layer can independently fulfill a user's request without requiring other layers to provide necessary services and support. Cloud computing systems are divided into different levels of encapsulation based on the encapsulation of underlying hardware resources, thereby transforming resources into services, as illustrated in Figure 1.13.

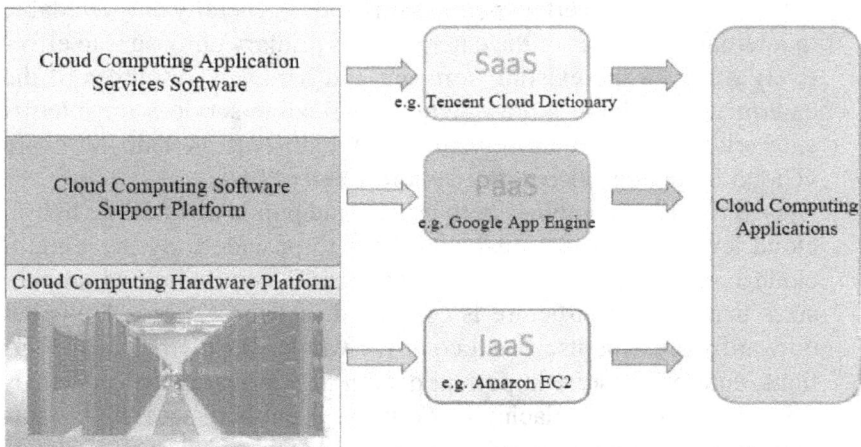

Fig. 1.13. Hierarchical service models and commercial representatives of cloud computing.

1.5. Major Deployment Models of Cloud Computing

The primary deployment models of cloud computing include public cloud, private cloud, hybrid cloud, and community cloud.

(1) *Public Cloud*: It is a cloud computing service targeted at the general Internet population. The audience for public cloud services encompasses all individuals within the Internet environment, and anyone can use the cloud computing services provided by simply registering and paying a certain fee. Users do not need to construct their own hardware, software, or other infrastructure, nor do they need to maintain it. They can access and obtain resources from anywhere, at any time, and through various means via the Internet. Currently, mainstream public cloud platforms include AWS, Microsoft Azure, GAE, Alibaba Cloud, Sina App Engine (SAE), and Baidu App Engine (BAE).

Amazon's AWS offers a vast array of cloud-based global products, including computing, storage, databases, analytics, networking, mobile products, developer tools, management tools, IoT, security, and enterprise applications. AWS provides a secure, reliable, and scalable cloud services platform that helps businesses or organizations rapidly develop their businesses and reduce IT costs.

(2) *Private Cloud*: It is a cloud computing platform targeted at enterprise internal use. Accessing cloud computing services provided by a private cloud requires certain permissions and is generally only available to internal employees of the enterprise. Its primary purpose is to effectively organize the existing software and hardware resources of the enterprise, providing more reliable and flexible services for internal enterprise use. Popular private cloud platforms include VMware vCloud Suite and Microsoft's System Center 2016.

(3) *Hybrid Cloud*: It combines both private and public clouds. The hybrid cloud leverages the advantages of both to provide truly meaningful cloud computing services for enterprises. The hybrid cloud emphasizes that the infrastructure is composed of two or more clouds but outwardly presents itself as a cohesive whole. During regular operations, enterprises store important data in their private clouds (such as financial data) while placing less critical or publicly accessible information in the public cloud. The combination of these two clouds forms a unified entity, which is the hybrid cloud. Units like banks, for example, may find it difficult for their internal private cloud systems

to meet user demands during peak access periods. At such times, they can integrate with the public cloud to address additional user requests. The hybrid cloud represents the direction of cloud computing development, effectively utilizing the IT infrastructure that enterprises have invested heavily in while addressing concerns such as data security brought about by public clouds. It is the optimal solution to prevent enterprises from becoming information silos.

The key tool for building a hybrid cloud is OpenStack, which can integrate heterogeneous resources from various cloud computing platforms to construct enterprise-grade hybrid clouds. This enables enterprises to flexibly customize various cloud computing services according to their needs. When building an enterprise cloud computing platform, using the OpenStack architecture is one of the more ideal solutions.

(4) *Federated Cloud*: It brings together the cloud infrastructure of multiple cloud computing service providers to offer users more reliable and cost-effective cloud services, primarily targeting public cloud platforms. An example is Content Delivery Network (CDN) services deployed on cloud platforms, and the data content stored by the system is geographically dispersed, while users are also distributed around the world. If a user in Country A requests data content located in Country B, the data will traverse multiple routes, increasing network latency. Federated cloud automatically migrates the requested data resources to cloud data centers closer to the user, thus enhancing the service quality of CDN.

1.6. Cloud Computing: Innovating Business Models

Cloud computing is the industrialized deployment and commercial operation of large-scale computing capabilities. It represents a new, commercialized computing and service model, where computing power is allocated and used on demand, much like water, electricity, or gas. The emergence of cloud computing has led to a significant transformation in business models: Consumers are shifting from "purchasing hardware and software products" to "purchasing information services". Through centralized remote computing resource pools, cloud computing provides powerful and cost-effective computing services to end users on demand.

The development of cloud computing mirrors the evolution of electricity over the past 100 years. In the early days of electricity application, each household generated its own power. However, as power generation technology advanced and the number of electrical devices increased, electricity gradually became a social public infrastructure. Consequently, farms, companies, and households began to close their own power generators and instead purchased electricity from power plants, transitioning from purchasing power generation equipment to purchasing power services, achieving the goal of resource conservation. Similarly, cloud computing can be viewed as an "information power plant" (from a business perspective, cloud computing is equivalent to an information power plant). Cloud computing, through data centers, provides various service capabilities such as computing power, storage, and networking to users. Not only hardware but also software migrates from terminals to the cloud, allowing users to directly access services without the need for software installation. An important goal of cloud computing is to transform computing power into a utility service like water and electricity, being available on demand. Hence, cloud computing is also referred to as "utility computing". Here, "utility" does not refer to usefulness or practicality but has a specific meaning, referring to public services similar to water, electricity, and gas. Therefore, "utility computing" can be interpreted as "public service computing".

Cloud computing aims to reduce construction and operation costs, improve IT systems and business elasticity, and expand new service and business models. It is a computing system composed of large-scale, low-cost computing units interconnected via networks, providing various computing services to users. According to a report released by the renowned market research company Gartner, the global cloud computing market maintains a steady growth trend. In 2023, the global cloud computing market size, encompassing IaaS, PaaS, and SaaS, reached $586.4 billion, with a growth rate of 19.4%. It is projected that by 2026, the global cloud computing market size will surpass $1 trillion.

1.7. Typical Cloud Computing Products

Cloud computing, as a widely encompassing and profoundly impactful technology, has gradually permeated various aspects of the information industry and other industries, profoundly altering the structural, technological, and product sales patterns of the information industry, thus deeply influencing people's lives. Chinese companies such as Huawei, ZTE, Tencent, Alibaba, Lenovo, and Inspur have successively put forward their own cloud computing strategic plans and have comprehensively deployed in both cloud computing technology and the market. The following are introductions to several main types of cloud computing products.

1.7.1. *Amazon's AWS*

Amazon was the first company to sell cloud computing as a service. Amazon's cloud computing products are collectively referred to as AWS, with its logo depicted in Figure 1.14.

AWS mainly consists of the following components: EC2, S3, Simple Queuing Service (SQS), and SimpleDB, providing computing and storage services to enterprises. Chargeable services include storage space, bandwidth, CPU resources, and monthly fees. In less than two years since its inception, Amazon has amassed as many as 440,000 registered users, including numerous enterprise-level users. Additionally, Amazon also offers content delivery service CloudFront, e-commerce service DevPay, and FPS service. In other words, Amazon currently provides developers with storage, computing, middleware, and database management system services. Through AWS, users can access a scalable IT infrastructure service set according to the needs of their business, gaining computing power, storage, and other services. With AWS, users can flexibly choose which development platform or programming model to use based on the characteristics of the problem being solved. Users only pay for what they use, without upfront costs or long-term commitments, making AWS one of the most efficient ways to deliver applications to customers. Through AWS, Amazon.com's global computing infrastructure can effectively support its retail business and transaction enterprises, serving as many as 15 billion retail transactions. With AWS, an e-commerce website can

Fig. 1.14.   Logo of AWS.

easily adapt to unpredictable demands: A pharmaceutical company can rent computing power to perform large-scale simulations; a media company can provide virtually unlimited storage for videos, music, and more; and an enterprise can deploy bandwidth services that suit its changing business needs flexibly.

As a former e-commerce retail enterprise primarily focused on selling books, Amazon had to purchase a large amount of IT equipment to cope with sales peaks like the "Christmas rush" when designing and planning its own IT system architecture. However, these devices remained idle most of the time. Therefore, in July 2002, Amazon launched the free Amazon E-commerce Service, allowing retailers to place their products in the Amazon online store, store product prices, customer reviews, and other information for backend management. In this way, Amazon was not only selling books but also leveraging its expertise in e-commerce website construction to offer services to other enterprises as a bundled product, with storage servers and bandwidth charged according to capacity and CPU usage charged based on usage duration and workload. To address reliability, flexibility, security, and other issues in these leased services, Amazon continuously optimized its technology. In 2006, AWS began to provide professional cloud computing services to Amazon and other enterprises in the form of web services. By 2017, Amazon's cloud computing service AWS had reached a revenue scale of $17.5 billion. Currently, Amazon offers users a comprehensive set of services, including elastic computing, storage services, databases, and applications, helping enterprises reduce IT infrastructure investment costs and maintenance costs. Amazon AWS has become one of the leading cloud computing infrastructure service providers in the global market in terms of market share.

1.7.2. *Windows Azure platform*

Microsoft closely followed the development trend of cloud computing and launched the Windows Azure operating system in October 2008. Azure represents another transformative shift for Microsoft, similar to when Windows replaced DOS. It involves building a brand-new cloud computing platform on the Internet architecture, enabling Windows to truly extend from PCs to the cloud.

The Windows Azure Platform, developed by Microsoft, is a cloud computing operating system designed to provide the operating system,

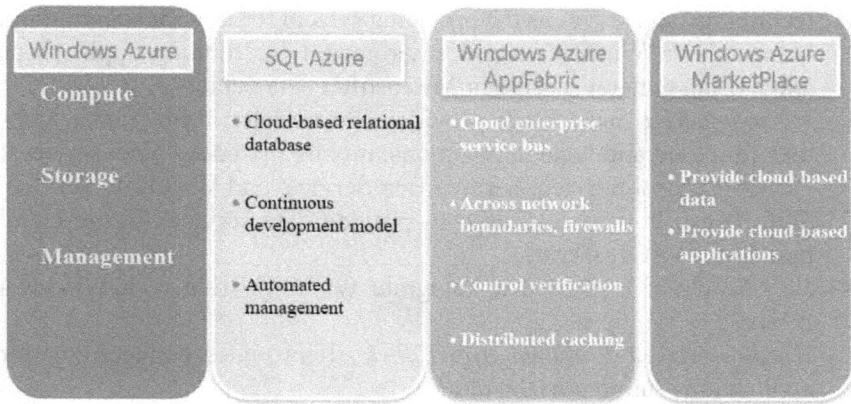

Fig. 1.15.   Functional diagram of the Windows Azure Platform.

basic storage, and management platform required for cloud-based services. It marks Microsoft's first step into cloud computing and is a part of Microsoft's online service strategy. It falls under the category of PaaS cloud computing service model, and its various functionalities are illustrated in Figure 1.15.

The Windows Azure Platform was announced by Microsoft's Chief Software Architect, Ray Ozzie, on October 27, 2008, at Microsoft's annual Professional Developers Conference, with its community preview version. It officially began commercial operation (RTM Release) in February 2010. Its seven data centers are located in Chicago, San Antonio, and Texas in the United States, Dublin in Ireland, Amsterdam in the Netherlands, Singapore, and Hong Kong in China. In March 2014, Microsoft's public cloud Azure was officially launched in China. In November 2018, the Customer Engagement Fabric (CEF) was introduced as a preview version of the user connection service, offering basic services for app and user interaction connections, including multi-channel notification services, third-party login, and aggregated payment functionality.

Microsoft's Windows Azure Platform is a collection of cloud technologies, each providing a range of services for application developers. It includes the following parts:

(1) *Windows Azure*: At the lowest level of the computing platform, Windows Azure is the core of Microsoft's cloud computing

technology and serves as the operating system for its cloud computing. It provides a Windows-based environment for running applications and storing data on servers in Microsoft's data centers.

(2) *Microsoft.NET Services*: It provides common basic functional modules for cloud and local applications, mainly including three services: Access Control Service, Service Bus Service, and Workflow Service.

(3) *SQL Azure*: It is mainly used to provide data services in the cloud based on an SQL server.

(4) *Live Services*: It is used to integrate Windows Live with Windows Azure.

(5) *Windows Azure Platform App Fabric*: It provides connectivity for applications running in the cloud.

1.7.3. *IBM Blue Cloud solution*

IBM is a leader in commercial data computing and traditional supercomputing. In the field of cloud computing, IBM is a comprehensive provider of hardware, software, and services. IBM views cloud computing as a strategic priority and has established 13 cloud computing centers worldwide. With numerous successful cases, IBM has assisted many clients in China in deploying cloud computing centers. IBM can help enterprises establish internal private clouds and also provide external public cloud services. IBM has invested heavily in research and development of cloud computing technology and plans to continue investing heavily in the next two to three years to support cloud computing development, thereby establishing a cluster of supercomputers that operate like a single computer.

In 2007, with IBM and Google both naming some of their projects as cloud computing, the concept of cloud computing began to rapidly spread. That same year, IBM launched the "Blue Cloud" initiative, becoming one of the first traditional IT companies to announce a cloud computing strategy. In June 2008, IBM established the Greater China Cloud Computing Center in Beijing. This center provides infrastructure design and implementation services of cloud computing centers, high-skilled manpower support of cloud computing, training of next-generation data center services, and rapid deployment and implementation of cloud computing concepts and pilot runs. In early 2011, IBM restructured its software, hardware, and services divisions to tackle the challenge of cloud computing, establishing the IBM Cloud Computing Division. By 2015, cloud computing became one of IBM's core development plans to facilitate IBM's ongoing transformation and create higher value. In 2016, IBM significantly expanded its public cloud data centers, opening new infrastructure in Norway, South Africa, and the United Kingdom. Now, the IBM Cloud is accessible from over 50 locations across six continents worldwide. Currently, IBM Cloud Computing ensures seamless integration into both public and private cloud environments. Its infrastructure is secure, scalable, and flexible, providing customized enterprise solutions, making IBM a leader in the hybrid cloud market. The IBM Blue Cloud Solution is depicted in Figure 1.16.

The "Blue Cloud" solution consists of the following components:

(1) Hardware and software resources that need to be included in the cloud computing center. Hardware may include machines with x86 or Power PC architecture, storage servers, switches, routers, and other network devices. Software may include various operating systems, middleware, databases, and applications, such as AIX, Linux, DB2, WebSphere, Lotus, and Rational.
(2) Blue Cloud management software and IBM Tivoli management software. The Blue Cloud management software, developed by IBM Cloud Computing Center, is specifically designed to provide cloud computing services.
(3) Blue Cloud consulting services, deployment services, and customization services. The Blue Cloud solution can be further customized according to the specific needs and application scenarios of customers, enabling integration with existing customer software and hardware.

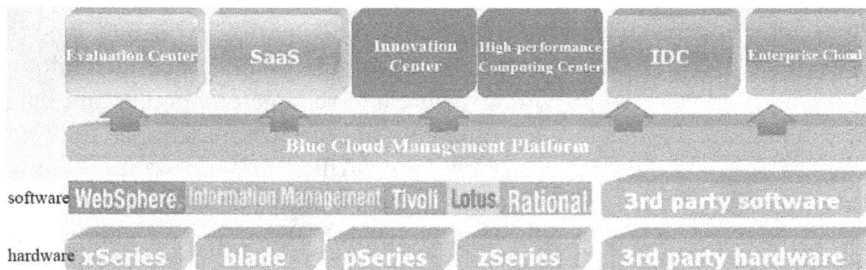

Fig. 1.16.   IBM Blue Cloud solution diagram.

This solution can automatically manage and dynamically allocate, deploy, configure, reconfigure, and reclaim resources, as well as automatically install software and applications. Blue Cloud can provide users with virtual infrastructure. Users can define the composition of virtual infrastructure, such as server configuration, quantity, storage type and size, and network configuration. Users submit requests through a self-service interface, and the lifecycle of each request is maintained by the platform. This solution can support 6+1 application scenarios, each with different component configurations and combinations of software and hardware, hence referred to as the 6+1 solution.

In the field of cloud computing, IBM has provided the following successful cases:

(1) Wuxi Cloud Computing Center—Software Development and Testing Cloud: IBM collaborated with Wuxi City to establish its first cloud computing center in China, aimed at accelerating its software outsourcing business and providing IT services to software developers in the region, gradually transitioning toward a service-oriented economy.
(2) i-Tricity Cloud Computing Center—IDC Cloud: i-Tricity is a cloud computing service provider based in Amsterdam, Netherlands. It selected IBM to establish the Blue Cloud computing center, providing 24/7 cloud computing services to companies in Belgium, the Netherlands, and Luxembourg.
(3) Vietnam Technology & Telecommunication Association (VNTT) collaborated with IBM to establish a telecommunications cloud base in Phu Yen Province, Vietnam.

1.7.4. *Alibaba Cloud*

As a unicorn enterprise in the field of cloud computing, Alibaba Cloud is a leading global cloud computing and artificial intelligence technology company, dedicated to providing secure and reliable computing and data processing capabilities through online public services, making computing and artificial intelligence accessible to all. Alibaba Cloud serves leading enterprises in various fields such as manufacturing, finance, government affairs, transportation, healthcare, telecommunications, and energy. Its clients include large enterprises such as China Unicom, 12306, Sinopec, PetroChina, Philips, and BGI, as well as heavyweight Internet products

⊂-⊐ 阿里云

Fig. 1.17. Logo of Alibaba Cloud.

like Weibo and Zhihu. Alibaba Cloud has maintained a good operational record in extremely challenging application scenarios such as the Tmall "Double Eleven" global shopping festival and 12306 Spring Festival ticket purchasing. Additionally, Alibaba Cloud has deployed efficient and energy-saving green data centers worldwide, using clean computing to provide a continuous power supply for the interconnected world. Currently, Alibaba Cloud services are available in regions including China, Singapore, the United States, Europe, the Middle East, Australia, and Japan. The logo of Alibaba Cloud is shown in Figure 1.17.

Among them, Elastic Compute Service (ECS) is a basic cloud computing service provided by Alibaba Cloud. Users can use ECS instances as conveniently and efficiently as they use resources like water, electricity, and gas. Users do not need to purchase hardware equipment in advance. Instead, they can create the required number of ECS instances anytime according to business needs. During usage, users can expand disk capacity and increase bandwidth as their business grows. If ECS instances are no longer needed, resources can be released at any time to save costs.

As shown in Figure 1.18, all resources involved in ECS are listed, including instance types, block storage, images, snapshots, bandwidth, and security groups. Users can configure ECS resources through the Alibaba Cloud console or the Alibaba Cloud app.

Alibaba Cloud's ECS offers a variety of block storage product types, including elastic block storage based on distributed storage architecture and local storage based on physical machine local disks.

(1) *Elastic Block Storage*: Elastic Block Storage is Alibaba Cloud's data block-level random storage provided for ECS instances. It features low latency, persistence, and high reliability. Utilizing a three-replica distributed mechanism, it ensures 99.99% data reliability for ECS instances. It can be created, released, or expanded at any time.

(2) *Local Storage*: Also known as local disks, local storage refers to the local hard disks mounted on the physical machine (host) where the ECS cloud server resides. It serves as temporary block storage

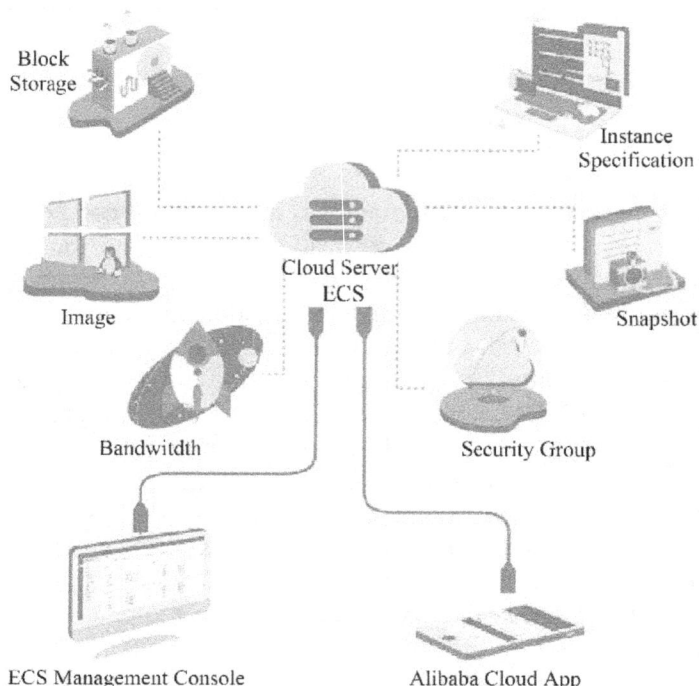

Fig. 1.18.   All resources involved in ECS.

designed for scenarios with extremely high storage I/O performance requirements. This type of storage provides block-level data access capabilities for instances with low latency, high random I/O per second (IOPS), and high throughput I/O capabilities.

Alibaba Cloud currently offers three main data storage products: Block Storage, Network Attached Storage (NAS), and Object Storage Service (OSS). Here are the differences between them:

(1) *Block Storage*: Block Storage is provided by Alibaba Cloud for ECS cloud servers. It offers high performance, low latency, and supports random read and write operations, similar to using physical hard disks. It can be used for storing data in most general business scenarios.

(2) *OSS*: OSS can be understood as a massive storage space, which is most suitable for storing large volumes of unstructured data generated on the Internet, such as images, short videos, and audio files. Users can access data in object storage at any time and from anywhere using APIs. It is commonly used for building Internet business websites, separating dynamic and static resources, CDN acceleration, and other business scenarios.

(3) *NAS*: Similar to object storage, NAS is suitable for storing large volumes of unstructured data. However, users need to access this data through standard file access protocols, such as the Network File System (NFS) protocol for Linux systems and the Common Internet File System (CIFS) protocol for Windows systems. Users can set permissions to allow different clients to access the same file simultaneously. NAS is suitable for file sharing between enterprise departments, broadcast nonlinear editing, high-performance computing, Docker, and other business scenarios.

1.8. New Developments in Cloud Computing Technology

Cloud computing, as an emerging technology, has experienced rapid development, fundamentally changing the way people work in many fields and transforming traditional software enterprises. This section discusses some of the most prominent new technologies in the current stage of cloud computing development.

1.8.1. *Software defined storage (SDS)*

SDS does not yet have a precise definition, but in simple terms, it refers to applications running on any storage that can automatically operate under user-defined policies. In data centers, resources such as servers, storage, networks, and security can be defined and automatically allocated using SDS. The core of SDS is storage virtualization technology. Software-defined data centers support constantly changing business needs through existing resources and applications, thus achieving IT flexibility. The core idea is to pool resources—processors, networks, storage, and possible middleware—in such a way that atomic units of computation can be generated and easily allocated or withdrawn based on business process requirements. SDS can be installed on commercial resources (x86 hardware, virtual machine monitors, or clouds) and existing computing hardware with any storage software stack.

SDS involves extracting typical storage controller functions from storage hardware and implementing them in software. These functions include volume management, RAID, data protection, snapshots, and replication. Since the functions of the storage controller are implemented by SDS, this functionality can reside in any part of the infrastructure, creating a truly converged architecture and a simpler, more scalable architecture. The main advantage of SDS is its significant cost reduction compared to traditional storage and its tight integration with existing virtual architectures.

SDS provides agility and rapid scalability while implementing workload separation. With the gradual popularity of SDS hybrid cloud trends, SDS has become a mainstream technology and is evolving into a specific architectural approach. Currently, many mainstream vendors are able to provide SDS solutions. The ZFS software stack is a popular SDS solution, along with other commercial software stacks such as Nexenta. Some proprietary SDS software has also emerged, including GreenBytes, VMware's acquired Virsto SVC, HP's LeftHand series (now StoreVirtual VSA), and Inspur's AS13000.

1.8.2. *Hyper-Converged Infrastructure (HCI)*

HCI, also known as hyper-converged architecture, refers to the concept that, in a single set of unit equipment (x86 server), there are not only resources and technologies such as computing, networking, storage, and server virtualization but also other elements such as cache acceleration, data deduplication, online data compression, backup software, and snapshot technology. And multiple units can be aggregated over a network to achieve modular, seamless horizontal scalability (scale-out), forming a unified resource pool.

HCI resembles the large-scale HCI patterns seen in the backends of companies like Google and Facebook, offering optimal efficiency, flexibility, scalability, cost-effectiveness, and data protection for data centers. Hyper-converged architecture integrates virtualization technology and storage into the same system platform. In simple terms, it involves running virtualization software (Hypervisor) on physical servers and providing distributed storage services for virtual machines through the virtualization software. Distributed storage can run inside virtual machines on the virtualization software or be integrated as a module with the virtualization software. Broadly, in addition to virtualized computing and storage, hyper-converged architecture can also integrate networking and other platforms and services.

From a storage perspective, HCI is part of SDS. HCI belongs to the data layer and has online horizontal scalability, making it highly suitable for cloud environments. However, to achieve instant delivery, dynamic scaling, and online adjustments of storage resources required for cloud environments, a storage policy SDS is also needed at the control layer. The development of SDS relies on the implementation and vigorous development of hyper-converged architecture to take shape.

Currently, typical vendors representing hyper-convergence include Nutanix, VMware, SmartX, and Maxta. Since storage is the core of hyper-convergence, these vendors have implemented distributed storage solutions tailored for virtualization scenarios, such as Nutanix's NDFS, VMware's vSAN, SmartX's ZBS, and Maxta's MxSP.

1.8.3. *Software Defined Data Center (SDDC) and DevOps*

The Software Defined Data Center (SDDC) is a data management approach that abstracts computing, storage, and networking resources through virtualization and offers them as services. SDDC enables customers to achieve more flexible, rapid business deployment, management, and implementation at a lower cost. To facilitate this, SDDC employs automation and orchestration software to centrally manage virtualized resources and automate operational and allocation workflows. SDDC boasts three main advantages: agility, providing faster, more flexible business support and implementation, and optimizing and changing software development models; elasticity, offering dynamic scalability (both horizontally and vertically) according to business needs; and cost-efficiency, avoiding redundant hardware investments and resource waste through software implementation.

The architecture of the SDDC can be divided into three logical layers: the Physical Layer, the Virtualization Layer, and the Management Layer. Together, they provide a unified system that offers enterprises a more flexible and cost-effective way to operate data centers.

(1) *Physical Layer*: The physical layer of the SDDC architecture includes computing, storage, and networking components to support the storage and processing of enterprise data. The computing components consist of multiple server nodes combined in a cluster architecture, providing processing and storage resources to support data operations. Storage components may comprise various storage types such as SAN, NAS, or DAS, including HDDs and SSDs. The networking components of the SDDC architecture include physical hardware for communication between computing and storage resources and to safeguard enterprise data. Hardware for network components includes switches, routers, gateways, and any other necessary components to support SDDC communication in a clustered architecture.

(2) *Virtualization Layer*: Virtualization is the key to SDDCs, and the virtualization layer consists of software used to abstract underlying resources and provide them as integrated services. At the core of the virtualization layer is the Hypervisor, which presents resources as virtualized components.

Compute virtualization is based on server virtualization technology, which separates processor and memory resources from physical servers, forming a pool of logical computing components that enhance resource utilization. Applications rely entirely on virtualized processor and memory resources. Storage virtualization abstracts underlying physical devices and virtualizes storage resources into logical resource pools. Similar to SDS, storage virtualization abstracts details of underlying hardware, allowing each application to access the necessary storage resources without affecting other applications. Network virtualization separates available resources from underlying hardware, making physical bandwidth available as independent channels that can be dynamically allocated or reallocated to specific workloads in real time.

(3) *Management Layer*: Virtualizing physical resources is only part of the SDDC architecture; its infrastructure also includes a management layer capable of task orchestration and automation. The management layer includes monitoring, alerting, and scheduling capabilities so that administrators can supervise operations, maintain performance, and conduct advanced analytics. Additionally, this layer integrates with built-in security and data protection mechanisms in the SDDC architecture. The management layer also provides business logic, translating application requirements and requests into API instructions for executing

orchestration and automation operations. APIs enable management and virtualization software to configure and manage resources and address policy enforcement and service level agreements.

DevOps (a combination of Development and Operations) refers to a set of processes, methods, and systems aimed at facilitating communication, collaboration, and integration between development (application/software engineering), technical operations, and quality assurance (QA) departments. DevOps is a culture, method, or convention that emphasizes communication and collaboration between software developers (Dev) and IT operations technicians (Ops). It accelerates and makes software development, testing, and deployment more frequent and reliable through automated processes for "software delivery" and "architecture changes". The emergence of DevOps is due to the software industry's growing recognition that DevOps must work closely together to deliver software products and services on time.

1.8.4. *Rise of Hybrid Cloud services*

Hybrid cloud combines public and private cloud, emerging as a primary model and direction for cloud computing in recent years. Private clouds primarily target enterprise users who, for security reasons, prefer to store data in private clouds. However, they also seek access to the computing resources of public clouds. In this scenario, hybrid clouds are increasingly adopted, blending and matching public and private clouds to achieve optimal results. This personalized solution meets the dual objectives of cost-effectiveness and security.

Hybrid clouds offer many essential features that benefit enterprises of all sizes. These new features enable businesses to expand their IT infrastructure in unprecedented ways using hybrid clouds. Adopting hybrid clouds has several advantages:

(1) *Cost Reduction*: Cost reduction is one of the most appealing features of cloud computing and a significant factor driving enterprise management to consider adopting cloud services. Upgrading IT infrastructure can be costly, often requiring the purchase of additional servers, storage, and even the construction of new data centers. Hybrid clouds help reduce these costs by leveraging "pay-as-you-go" cloud

computing resources to lower or eliminate the need to purchase local IT equipment.

(2) *Increased Storage and Scalability*: Hybrid cloud provides an economically efficient way for enterprises to expand storage. The cost of cloud storage is much lower compared to equivalent local storage costs, making it an ideal choice for backup, virtual machine replication, and data archiving.

(3) *Enhanced Agility and Flexibility*: One of the greatest benefits of hybrid cloud is flexibility. It allows enterprises or individuals to migrate resources and workloads from on-premises to the cloud and vice versa. Hybrid cloud empowers developers to easily "spin up" new virtual machines and applications without the assistance of IT operations personnel. Enterprises can also leverage the elastic scalability of hybrid cloud to move some applications to the cloud to handle peak business demands.

1.8.5. *Edge computing*

Different organizations provide different definitions for edge computing. Professor Shi Weisong and others from the Department of Computer Science at Wayne State University in the United States define edge computing as "a new computing paradigm that performs computations at the edge of the network. In edge computing, downstream data at the edge represents cloud services, while upstream data represents the Internet of Things services". The Edge Computing Industry Alliance defines edge computing as "a development platform that integrates network, computing, storage, and application core capabilities on the edge side close to the physical or data source. It provides edge intelligence services nearby to meet the key requirements of industries in agile connections, real-time businesses, data optimization, application intelligence, security, and privacy protection".

Therefore, edge computing is a new computing paradigm that provides integrated computing, storage, and network resources for applications close to the physical or data source at the network edge. Additionally, edge computing is an enabling technology that meets key industry requirements such as agile connections, real-time business, data optimization, application intelligence, security, and privacy protection by providing these resources at the network edge.

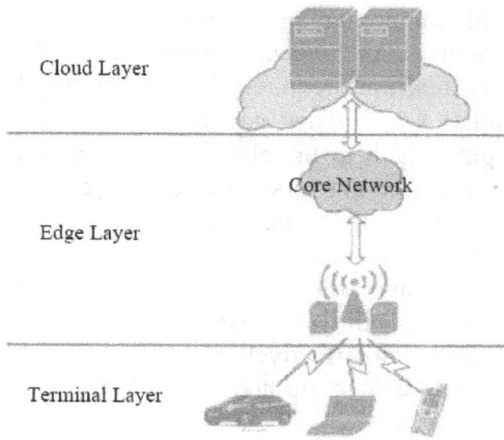

Fig. 1.19. Architecture of an edge cloud.

(1) *Architecture of Edge Computing*: Edge computing extends cloud services to the network edge by introducing edge devices between terminal devices and the cloud. The architecture of edge computing includes the terminal layer, edge layer, and cloud layer. The composition and functions of each layer in the architecture of edge computing are briefly introduced in the following. The architecture of edge computing is shown in Figure 1.19.

- *Terminal Layer*: The terminal layer is the closest layer to end users. It consists of various Internet of Things (IoT) devices, such as sensors, smartphones, smart vehicles, smart cards, and card readers. To prolong the service life of terminal devices, it is advisable to avoid running complex computing tasks on them. Therefore, terminal devices are only responsible for collecting raw data and uploading it to the upper layers for computation and storage. The terminal layer is connected to the upper layer primarily through cellular networks.

- *Edge Layer*: The edge layer is located at the edge of the network and consists of a large number of edge nodes, typically including routers, gateways, switches, access points, base stations, specific edge servers, etc. These edge nodes are widely distributed between terminal devices and the cloud layer, such as in cafes, shopping centers, bus stations, streets, and parks. They are capable of computing and storing the data uploaded by the terminal devices. As these edge nodes are closer to users, they can meet users' real-time requirements for running latency-sensitive applications. Edge nodes can also preprocess collected data before uploading it to the cloud, thereby reducing transmission traffic on the core network. The edge layer is connected to the cloud layer primarily through the Internet.

- *Cloud Layer*: The cloud layer consists of multiple high-performance servers and storage devices. It has powerful computing and storage capabilities to execute complex computing tasks. Cloud modules can effectively manage and schedule edge nodes and cloud data centers through control policies, providing users with better services.

(2) *Advantages of Edge Computing*: The edge computing model involves migrating some or all computational tasks from traditional cloud computing centers to locations near data sources. Compared to traditional cloud computing models, the edge computing model offers advantages such as real-time data processing and analysis, high security, privacy protection, strong scalability, location awareness, and low traffic.

- *Real-time Data Processing and Analysis*: By moving some or all computational tasks from cloud computing centers to the network edge, data can be processed on edge devices rather than externally in data centers or the cloud. This improves data transmission performance, ensures real-time processing, and reduces the computational load on cloud computing centers.
- *High Security*: Traditional cloud computing models are centralized, making them vulnerable to Distributed Denial-of-Service (DDoS) attacks and power outages. The edge computing model distributes processing, storage, and applications between edge devices and cloud computing centers, enhancing security and reducing the risk of single points of failure.
- *Protection of Privacy Data and Enhanced Data Security*: Edge computing involves processing more data on local devices rather than uploading it to cloud computing centers. Therefore, edge computing can also reduce the volume of data that is actually at risk. Even if a device is attacked, only locally collected data are compromised, while the cloud computing center remains unaffected.

- *Scalability*: Edge computing offers a more cost-effective scalability path, allowing companies to expand their computing capabilities through a combination of IoT devices and edge data centers. Using IoT devices with processing capabilities also reduces expansion costs, as new devices do not impose significant bandwidth demands on the network.
- *Location Awareness*: Edge distributed devices share information using low-level signaling. The edge computing model receives information from edge devices within the local access network to discover device locations. For example, in navigation, terminal devices can transmit relevant location information and data to edge nodes for processing, and the edge nodes make judgments and decisions based on existing data.
- *Low Traffic*: Data collected by local devices can be locally analyzed or preprocessed on local devices, eliminating the need to upload all locally collected data to cloud computing centers, which can reduce the traffic entering the core network.

1.8.6. *Distributed Cloud*

In recent times, Distributed Cloud has emerged as a buzzword in the cloud computing market. Gartner, the consulting firm, defines Distributed Cloud as the distribution of public cloud services (typically including necessary hardware and software) to different physical locations (i.e., the edge), while ownership, operation, governance, and ongoing development of the services remain the responsibility of the original public cloud provider. In simple terms, Distributed Cloud refers to clouds distributed across different geographical locations and represents the latest evolution of cloud computing. Currently, cloud providers need to offer more up-to-date solutions for industry users to achieve lower latency, on-demand customization, high flexibility, consistent experiences, and robust security, highlighting the value of local or distributed clouds. For government and enterprise users, the benefits of Distributed Cloud are becoming increasingly apparent, as the cloud itself is inherently topical and story-driven, making it easier to arouse curiosity and novelty in the consumer market, garnering widespread attention and market value.

From the perspective of network evolution, operator requirements, and business development, constructing a multi-tiered Distributed Cloud

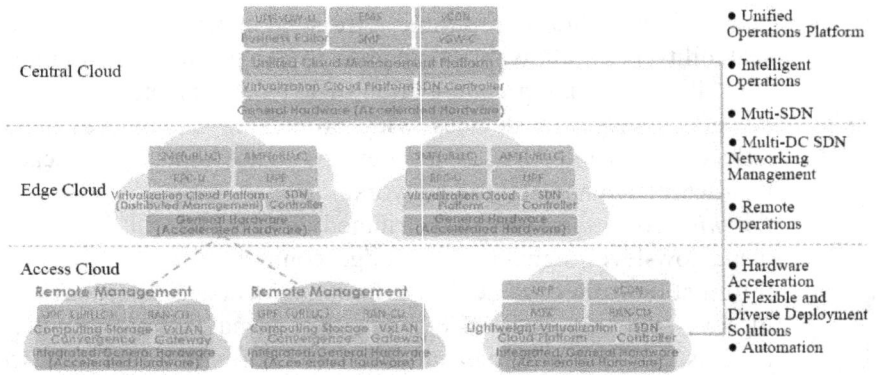

Fig. 1.20.   Architecture of Distributed Cloud.

to meet the characteristics of various business layers and deployment location requirements has become the trend in the evolution of cloud infrastructure. Based on current business development, Distributed Cloud can be divided into three tiers, as illustrated in Figure 1.20.

(1) *Central Cloud*: The central cloud primarily hosts control/management elements and centralized media-facing elements to improve management efficiency. Central clouds are generally large in scale and have complex operations/configuration, thus demanding strong requirements for unified management, automated management, and network automation configuration.

(2) *Edge Cloud*: The Edge Cloud primarily hosts distributed deployment of user-facing/media-facing elements to meet user experience requirements. It is mainly used for traffic offloading and requires strong network performance (primarily throughput). Therefore, various acceleration technologies become essential choices for it.

(3) *Access Cloud*: The access cloud primarily hosts access-oriented network elements and high real-time requirement elements, such as AR/VR and vehicle networking. Positioned closest to users, the access cloud considers its business and coverage scope, characterized by small scale, large quantity, heterogeneous deployment, and unmanned operation. It also demands high performance (throughput, latency), presenting challenges in management and performance enhancement.

1.8.7. *Xinchuang Cloud*

On September 30, 2020, the first national cloud platform project based on the Xinchuang Cloud architecture was officially delivered online, marking the debut of China's deployment of the Xinchuang Cloud. The inaugural Xinchuang Cloud project, deployed using the domestically developed general-purpose cloud operating system AnLinux OS and AnLinux CloudSuite cloud management suite, facilitated comprehensive and efficient digital operation and management across all sectors of the government. This project represents a significant practical step in promoting China's independent and controllable information technology. To develop Xinchuang, the first priority was to solve the security issue. For a long time, China's critical information systems and key infrastructure have mostly relied on foreign core information technology products and essential services, leaving China's data security in a state of uncertainty. The crackdown by the United States on Huawei and ZTE serves as a warning bell, emphasizing the critical need to develop independent and controllable technologies to reverse the situation of being constrained by core technologies. The establishment of the Information Technology Application Innovation Committee regards the Xinchuang Cloud as a national strategy, building on the foundations of the information technology industry to construct its own IT industry standards and ecosystem. This enables IT products and technologies to be controllable, researchable, developable, and producible domestically, thereby creating a domestically controlled information technology industry ecosystem.

So, what is the Xinchuang Cloud? What is its relationship with Xinchuang, public cloud, and private cloud? What value will its development bring?

Obviously, the Xinchuang Cloud is, first and foremost, a cloud, representing a type of cloud computing with all the basic features of cloud technology. Public and private clouds are commonly distinguished based on scalability and cost considerations, and the Xinchuang Cloud shares these characteristics. However, its name "Xinchuang" emphasizes its key feature of independent security and control. The Xinchuang Cloud is a cloud platform independently developed on the foundations of domestically produced CPUs and operating systems, emerging against the backdrop of innovative information technology applications.

In the entire Xinchuang industry chain system, there are four major domains: IT infrastructure, basic software, application software, and security. Cloud computing, located in the IT infrastructure layer, is the core support of the new generation of information infrastructure. Therefore, the Xinchuang Cloud, which belongs to the cloud, can be said, to a certain extent, to determine whether the development cornerstone of the Xinchuang industry is stable and enduring.

From the periphery, Xinchuang Cloud needs to carry the underlying hardware and software infrastructure, including chips, complete machines, and operating systems, while also supporting new-generation enterprise applications such as big data, artificial intelligence, IoT, and 5G above. In the entire Xinchuang industry chain system, Xinchuang Cloud plays an important role in linking up and down, running through the ecosystem.

Although China lags behind foreign countries in core technology research and development in many aspects, in the field of cloud computing, China has basically started at the same time as foreign countries, and the development ecosystem is not inferior to that of foreign companies. Shortly after Xinchuang Cloud was proposed, a group of enterprises relying on government agencies, large Internet companies, and self-rising enterprises invested in its research and development.

With the close promotion of national policies, China's IT field already possesses a group of world-class basic software and hardware talents. Coupled with the urgent need for enterprise digital transformation, China has a huge market and a large number of users, all of which provide favorable conditions and broad development space for Xinchuang Cloud.

1.8.8. *Security becomes critical*

Cloud security, also known as cloud computing security, refers to a collection of strategies, technologies, and controls used to protect cloud computing data, applications, and related infrastructure. It is a subfield of computer security and network security, or more broadly, a subset of information security. Cloud security can be divided into two categories: one is the security of user data privacy, and the other is security against traditional internet and hardware devices.

Cloud computing has become a priority service promoted by governments around the world due to its cost savings, easy maintenance, and flexible configuration. However, the adoption of cloud computing services by users and government departments also poses challenges to the security of their sensitive data and critical business operations. Therefore, cloud security has become crucial for the development of cloud computing. Cloud security technologies have also rapidly evolved accordingly.

Cloud security aims to protect the information security of customers, enterprises, and government departments. Therefore, cloud security technologies first need to understand the needs of customers and provide solutions tailored to these needs, such as full disk or file-based encryption, customer key management, intrusion detection/prevention, Security Information and Event Management (SIEM), log analysis, two-factor authentication, and physical isolation.

Cloud security technology has always been committed to achieving secure and efficient management of cross-domain systems. Several effective security standards have been used to ensure the security of cloud computing, including Security Assertion Markup Language (SAML), Services Provisioning Markup Language (SPML), eXtensible Access Control Markup Language (XACML), and Web Service Security (WS-Security).

When designing a cloud computing platform, measures to address cloud security need to be considered, mainly from the following three aspects: vulnerability scanning and penetration testing, which are mandatory for all PaaS and IaaS; cloud security technology configuration management, where the most important element is configuration management, such as software upgrades or patch management (In SaaS and PaaS environments, configuration management is handled by the cloud computing service provider.); cloud security technology controls, where the cloud computing service provider is responsible for the operation and maintenance of all cloud computing infrastructure, including virtualization technology, networks, storage, and related code, such as management interfaces and APIs. Therefore, evaluating and controlling the development practices of the vendor is also essential.

1.9. Current Status of China's Cloud Computing Industry

1.9.1. *Government promotion of cloud computing industry development*

In April 2020, the National Development and Reform Commission and the Cyberspace Administration of China jointly issued the "Implementation

Plan for Promoting the Development of the New Economy through Cloud Computing and Big Data". This plan encourages the exploration of next-generation digital technologies such as big data, artificial intelligence, cloud computing, digital twins, 5G, IoT, and blockchain in industries and enterprises with the necessary conditions, aiming to provide technical support for the digital transformation of enterprises.

In April 2022, the Ministry of Industry and Information Technology initiated the compilation of the "Implementation Guidelines for Enterprises to Move to the Cloud in 2022", aiming to deepen the action of enterprises moving to the cloud and promote high-quality cloud adoption by enterprises.

In October 2022, the General Office of the Communist Party of China Central Committee and the General Office of the State Council issued the "Opinions on Strengthening the Construction of High-skilled Talent Teams in the New Era", which emphasizes the full utilization of new generation information technologies such as big data and cloud computing. It calls for strengthening the informatization construction of talent work, establishing and improving high-skilled talent pools, enhancing theoretical research and achievement transformation of high-skilled talents, promoting the development of high-quality courses, textbooks, and faculty construction that meet the needs of high-skilled talent cultivation, and developing high-level talent training standards and integrated courses.

In December 2022, the Central Committee of the Communist Party of China and the State Council issued the "Outline of the Strategic Plan for Expanding Domestic Demand (2022–2035)", proposing to systematically lay out new infrastructure and accelerate the construction of information infrastructure. It calls for accelerating the construction of the IoT, industrial Internet, satellite Internet, and gigabit optical network, establishing a national integrated big data center system, laying out and constructing national hub nodes for big data centers, promoting extensive and deep application of artificial intelligence, cloud computing, etc., and promoting the integration and intelligent configuration of "cloud, network, and terminal" resources. It aims to enhance the service capability of the national wide-area quantum secure communication backbone network according to demand.

In April 2023, the Ministry of Industry and Information Technology, the Cyberspace Administration of China, the National Development and Reform Commission, and eight other departments issued the "Implementation Opinions on Promoting the Evolution and Innovative

Development of IPv6 Technology". This document encourages the integration and innovation of IPv6 with 5G, artificial intelligence, cloud computing, and other technologies, supporting enterprises to accelerate the application of innovative "IPv6+" technologies such as application-aware networks, and new IPv6 measurement in various network environments and business scenarios. Through the evolution and upgrade of IPv6 technology, it aims to promote the coordinated development of data centers, cloud computing, and networks, continuously improve the quality of network transmission between data centers, and enhance service experience.

1.9.2. *Rapid development of China's cloud computing industry*

According to the 47th Statistical Report on Internet Development in China, large-scale cloud service providers in China are currently at the forefront of the global market, with their revenue maintaining high-speed growth. In terms of market share, Alibaba Cloud has become the world's third-largest public cloud service provider, with a market share ranking just below Amazon and Microsoft.

According to the Cloud Computing White Paper (2023) released by the China Academy of Information and Communications Technology (CAICT), the cloud computing market in China has maintained rapid growth, with a market size reaching 455 billion yuan in 2022, an increase of 40.91% compared to 2021. Among them, the public cloud market size increased by 49.3% to 325.6 billion yuan, and the private cloud market increased by 25.3% to 129.4 billion yuan. Compared with the global growth rate of 19%, the cloud computing market in China is still in a period of rapid development, maintaining a relatively high resilience under the pressure of a sluggish economy. It is expected that by 2025, the overall market size of cloud computing in China will exceed 1 trillion yuan.

Li Wei, Deputy Director of the Cloud Computing and Big Data Research Institute of CAICT, commented on the White Paper, stating that with the deepening of the digital transformation process, the digital economy has gradually become an important driving force for national economic growth. At the same time, the rapid development of large-scale artificial intelligence models has triggered a two-way transformation of digital application usage and computing resource supply, and the value of cloud computing as the operating system of the digital world is being fully demonstrated. On the one hand, cloud computing will redefine the usage of computing resources downward, while on the other hand, it will define new interfaces for digital applications upward, giving rise to a new paradigm of computing services.

Meanwhile, based on long-term research observations of the industry and discussions with frontline experts, the Cloud Computing and Big Data Research Institute of CAICT has identified the following as the top ten keywords for cloud computing in 2023: modern application, multi-core single cloud, distributed cloud, low/no code, software engineering, system stability, cloud-native security, cloud optimization governance, SMEs migrating to the cloud, and supercomputing/smart computing services.

In terms of technology, the development of cloud computing in China exhibits the following four characteristics: Firstly, x86 servers are the mainstream choice for cloud computing hardware platforms, with hardware accounting for a relatively high proportion of overall platform revenue. However, with the increasing standardization of hardware equipment and the enhancement of software heterogeneity capabilities, it is expected that the revenue share of software and services markets will gradually increase. Secondly, domestic cloud computing service providers attach importance to participating in the establishment of open-source ecosystems while actively conducting independent research and development. Chinese cloud computing service providers such as Alibaba Cloud, Tencent, and Huawei have successively participated in open-source foundations such as the Linux Foundation and the Cloud Native Computing Foundation (CNCF). In 2018, they released independently developed cloud computing products such as "Feitian 2.0" and "Redis 5.0". Thirdly, although security issues have been highly valued by cloud computing service providers, security incidents still occur frequently, and the capability for security risk control urgently needs to be further strengthened. Lastly, the collaboration between edge computing and cloud computing will greatly enhance the timely processing capability, data storage capacity, and deep learning ability of massive data, thereby promoting the further development of the IoT.

In terms of application, cloud computing applications in China are accelerating penetration from the Internet industry to traditional industries such as government affairs, finance, and industry. Firstly, government cloud is the most mature area of cloud computing applications. Currently, more than 90% of provincial-level administrative regions and 70% of municipal-level administrative regions nationwide have established or are constructing government cloud platforms. Secondly, the financial industry is actively exploring cloud computing application scenarios. Due to the low system migration costs of small- and medium-sized enterprises and Internet financial institutions and the strong demand for cloud computing applications, they tend to transform existing business systems through cloud computing. Finally, industrial clouds are beginning to be applied in various links of the industrial chain. By integrating with technologies such as industrial IoT, industrial big data, and artificial intelligence, various links of the industrial chain, such as industrial research and development design, production manufacturing, marketing, and after-sales service, have begun to introduce cloud computing for transformation, forming a new format and application model of intelligent development.

In terms of policy, various local government departments, cloud platform service providers, and cloud enterprises are encouraged to cooperate in promoting mechanisms. Support is provided for industrial and information technology authorities at all levels to establish expert advisory committees for enterprise cloud migration. Efforts are made to guide and promote the cloud migration of enterprises, strengthen policy interpretation, popularize cloud migration knowledge, enhance enterprise cloud migration awareness and practical capabilities, and continuously expand the influence of enterprise cloud migration. Support is given to local industrial and information technology authorities to establish comprehensive public service platforms, providing various services such as information system planning consultation, scheme design, supervision, and training for enterprises. In-depth evaluations of cloud service capabilities and service credibility are carried out, promoting the improvement of cloud computing enterprise service levels and service quality. Active exploration of insurance models to provide protection for cloud-migrated enterprises is encouraged. Support for accelerating innovation and entrepreneurship in the cloud is provided. Various types of enterprises and entrepreneurs are supported to use cloud computing platforms as the basis to actively cultivate new formats and models, such as platform economy and sharing economy, using new technologies, such as big data, IoT, artificial intelligence, and blockchain. The formulation and implementation of evaluation standards for the effectiveness of enterprise cloud migration are promoted, gradually establishing an evaluation system for the effectiveness of enterprise cloud migration. Third-party organizations are supported to evaluate and statistically analyze the effects of cloud migration based on relevant standards, including cost savings, efficiency improvements, business upgrades, and innovation promotion, guiding enterprises to migrate to the cloud deeply. Typical cases and successful experiences of enterprise cloud migration are summarized and publicized, with increased promotion efforts to create benchmark enterprises in cloud migration, fully leveraging their exemplary and leading role and achieving large-scale enterprise cloud migration. The implementation of relevant requirements of the Cybersecurity Law of the People's Republic of China is ensured, promoting the establishment of sound cloud computing-related security management systems and improving cloud computing network security protection standards. Cloud platform service providers are guided and supervised to effectively fulfill their main responsibilities, safeguarding user information security and business secrets.

Exercises

(1) Briefly describe what cloud computing is.
(2) What are the characteristics of cloud computing?
(3) Please explain what IaaS, PaaS, and SaaS are.
(4) What are the infrastructure components of cloud computing, and what functions do they serve individually?

Chapter 2

Overview of Big Data Technologies

In distributed computing, the size and format of data often exceed the capabilities of typical database software in terms of gathering, storing, managing, and analyzing. Therefore, the emergence of new technologies is necessary to handle and analyze the current volume of data. This is where big data technology comes into play. This chapter primarily discusses the emergence of big data technology, including its characteristics, relevant technologies, and major applications, and finally introduces the relationship between cloud computing and big data.

2.1. Emergence of Big Data Technologies

The rapid development of information technology and its widespread application in various industries, coupled with the rapid expansion of the scale of industry application systems, has led to exponential growth in data. Big data, which can reach hundreds of TB or even tens to hundreds of PB, has far exceeded the processing capacity of traditional computing technology and information systems. This trend has driven the generation and rapid development of big data technology.

2.1.1. *Basic concepts of big data*

According to Wikipedia, big data refers to large or complex datasets that traditional data processing applications are inadequate to handle. It can also be defined as a large amount of structured and unstructured data from various sources. Typically, the volume of data in big data exceeds the processing capacity of traditional software within acceptable time limits.

As defined by Think Tank, big data refers to a collection of data whose content cannot be captured, managed, and processed with conventional software tools within a certain period of time. Big data technology refers to the ability to rapidly extract valuable information from various types of data. Technologies suitable for big data include Massively Parallel Processing (MPP) databases, data mining, distributed file systems, distributed databases, cloud computing platforms, the Internet, and extensible storage systems.

Baidu Encyclopedia defines big data as a collection of data sets that cannot be captured, managed, and processed within a certain time frame using conventional software tools. It encompasses massive, high-growth, and diverse information assets that require new processing methods to enhance decision-making, insight discovery, and process optimization capabilities.

In summary, big data refers to large, complex datasets that are difficult to handle through existing database management tools and traditional data processing methods. The scope of big data technology includes data collection, storage, search, sharing, transmission, analysis, and visualization.

2.1.2. *Causes of big data generation*

(1) *The Emergence of Big Data*: In the 21st century, with the comprehensive integration of computer technology into all aspects of social life, the widespread application of information technology has sparked technological innovation and business transformation. Numerous fields, such as the Internet (social networks, search engines, e-commerce), video websites, mobile Internet (microblogs, X), the Internet of Things (IoT), connected cars, GPS, medical imaging, security monitoring, finance (banking, stock markets, insurance), and telecommunications (calls, SMS), are generating vast amounts of data at an unprecedented pace. These data not only inundate the world with more information than ever before but also give rise to the widely known concept of "big data".

The emergence of big data technology is primarily driven by the increasing demand for network data analysis by Internet companies, as depicted in Figure 2.1. In the 1980s, a typical representative was

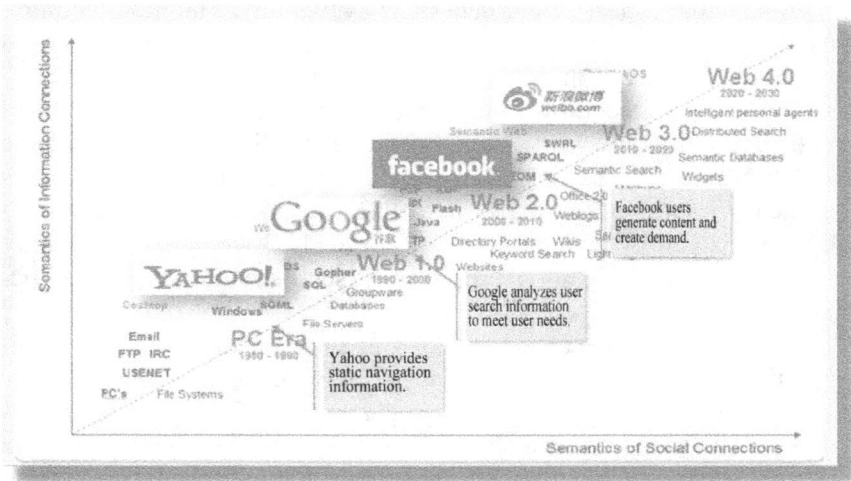

Fig. 2.1. Internet companies' demand for network data analysis.

Yahoo's "Directory" search database; in the 1990s, it was Google, which began using algorithms to analyze user search information to meet their actual needs. In the 21st century, Facebook emerged as a typical representative, not only meeting users' actual needs but also creating new demands as the advent of Web 2.0 transformed people from passive information recipients to active creators. After 2010, social networking sites such as YouTube, X, and microblogs appeared, generating massive amounts of videos, images, text, and short messages through these platforms. The growth rate of Internet-based data has become similar to Moore's law in the IT industry, which reveals the pace of technological advancement.

(2) *Availability and Derived Value of Big Data*: Since the onset of the IT era, humanity has amassed vast amounts of data, which continue to proliferate at an exponential rate. This has brought about two significant changes: Firstly, applications that were previously unattainable due to the lack of accumulated data are now feasible; secondly, the transition from an era of data scarcity to one of data abundance has presented new challenges and dilemmas in the processing and application of data, specifically in efficiently retrieving the necessary data from this vast pool, effectively processing it, and ultimately deriving valuable insights.

An essential aspect of big data is its availability. The more comprehensive the data used for analysis, the closer the resulting analysis is to reality, thus making it more usable. Data availability primarily encompasses methods for acquiring and integrating high-quality data, establishing theoretical frameworks for big data availability, approximate calculation and data mining of weakly available data, describing data consistency issues, automatically detecting consistency errors, automatically repairing entity integrity, and automatically detecting entity identity errors and entity recognition problems in semi-structured and unstructured data.

Continuing from the above scenario, in 2006, the data volume of individual users entered the era of terabytes (TB), with the hard drive capacity of personal computers upgrading from gigabytes (GB) to the TB specification. Globally, approximately 180 exabytes (EB) of new data were generated. By the year 2010, the global data volume reached the zettabyte (ZB) level, and in 2011, this number reached 1.8 ZB. IDC estimates that by 2022, the global storage of big data will reach 61.2 ZB, as projected by the China Industrial Research Institute. It is worth noting that the units of data volume increase successively from kilobytes (KB), megabytes (MB), gigabytes (GB), terabytes (TB), petabytes (PB), exabytes (EB), zettabytes (ZB), yottabytes (YB), brontobytes (NB), to geopbytes (DB). It is important to highlight that data measured in PB is considered big data.

Another important aspect of big data is the complexity of the data. Currently, 85% of the data belongs to unstructured and semi-structured data generated from social networks, the IoT, e-commerce, and other sources. Unstructured data refer to data with irregular or incomplete structures, without predefined data models, and is not easily represented using two-dimensional logical databases. This includes all formats of office documents, text, images, graphics, and audio/video information. Semi-structured data falls between fully structured data (found in relational and object-oriented databases) and completely unstructured data. XML and HTML documents are examples of semi-structured data; they are usually self-descriptive, with the structure and content intertwined without clear distinctions.

The structure of big data is becoming increasingly complex, far surpassing the capabilities of traditional methods and theories. Sometimes, even small data within big data, such as a single microblog post, can have a disruptive effect. Therefore, to make these new data structures and big data usable, new technologies and methods are required to collect, clean, analyze, and process current big data, extracting valuable knowledge from it.

Big data itself is challenging to utilize directly; only through processing can big data truly become valuable. Despite the aforementioned challenges, with the continuous growth of big data, it becomes evident that, through the adoption of new methods and technologies, these vast datasets are indeed usable and hold immense value.

Big data has the potential to create significant derived value across various domains, shifting the focus of future IT investments away from system building as the core and toward big data. The efficiency in handling big data gradually becomes vital to a company's vitality. The potential value of big data is illustrated in Figure 2.2.

From the graph, it can be observed that the government, real estate, healthcare, finance and insurance, public utilities, and consulting services industries possess relatively larger volumes of data compared to other sectors, particularly the government and real estate industries, which boast extensive datasets. The potential value of big data in the government, education services, and cultural entertainment industries is relatively moderate, with relatively lower development difficulty. The potential

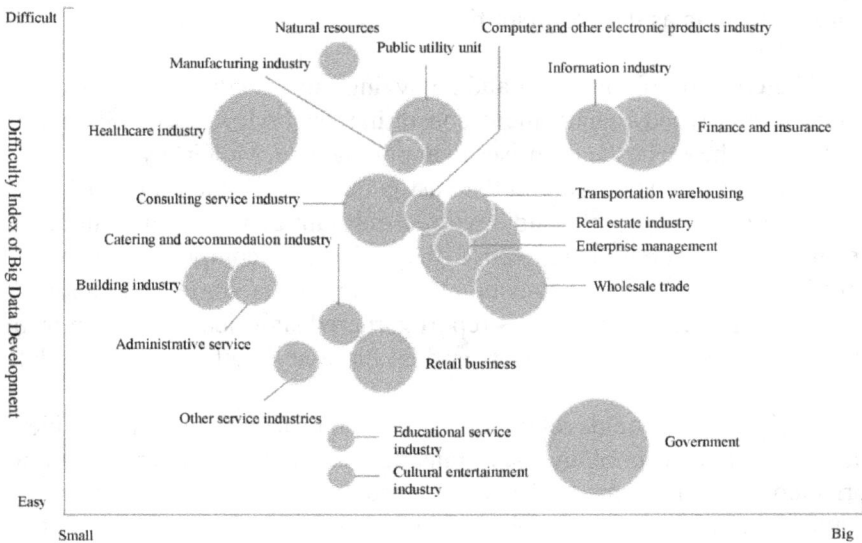

Fig. 2.2. Big Data value potential index.

value of big data in industries such as catering, real estate, consulting services, and retail falls in the middle range, with medium development difficulty. Industries like healthcare, natural resources, IT, finance and insurance, and public utilities exhibit greater potential value in big data, albeit with higher development difficulty.

2.1.3. *Introduction of the concept of big data*

Due to the emergence of massive unstructured and semi-structured data, conventional software methods are no longer capable of completing storage, management, and processing tasks within acceptable time frames. How to handle such data has become a crucial issue. In 2008, the journal *Nature* launched a special issue on big data, sparking attention from the academic and industrial sectors. Data has transformed into both the subject and tool of scientific research, prompting industries to contemplate, design, and implement scientific research based on data. Data is no longer merely the result of scientific research but has become the foundation of scientific inquiry.

Although "big data only became a popular term in the Internet industry from 2009, back in 1980, the renowned futurist Toffler enthusiastically praised big data as the "brilliant finale of the third wave" in his book, *The Third Wave*.

The concept of collecting and analyzing big data originated from the globally renowned management consulting firm McKinsey, which was also one of the earliest enterprises to apply big data. McKinsey recognized the potential business value in the vast personal information recorded on various online platforms and invested substantial human and material resources in research. In June 2011, they released a comprehensive report on big data, analyzing its societal impact, key technologies, and application areas in detail. McKinsey's report garnered significant attention from the financial sector, leading to a gradual increase in attention to big data across various industries.

Reflecting on the development of computer technology, it is evident that computer technology has transitioned from being computation-oriented to data-oriented, also more accurately termed as "data-centric computing". A data-centric approach necessitates the design and architecture of systems to revolve around data as the core. The description of this process is illustrated in Figure 2.3, which provides a chronological and

Data-oriented development

The government work report provides a definition of cloud

John McCarthy invented the LISP language   SUN proposed that the network is the computer

Cloud computing enters China

McCarthy first proposed the concept of cloud computing

Parallel Computing Grid Computing Google proposes cloud computing Fourth Paradigm

| 1958 | 1960 | 1984 | 1994 | 1995 | 2006 | 2008 | 2009 | 2012 |

The development of hardware technology and network technology

Large-Scale Integration

Invented WWW

Electronic tube: ENIAC   Transistor: IRADIC   TCP/IP   Internet development

Integrated circuit: IBM360   ARPAnet   China's first E-Mail   The first Web application   Mobile Internet

| 1946 | 1954 | 1964 | 1969 | 1970 | 1980 | 1986 | 1989 | 1993 | 2002 | 2005 |

Trends of Change: (1) From computation-centric   (2) From hardware-centric   (3) From centralized to
to data-centric   to network-centric   decentralized and then back to centralized

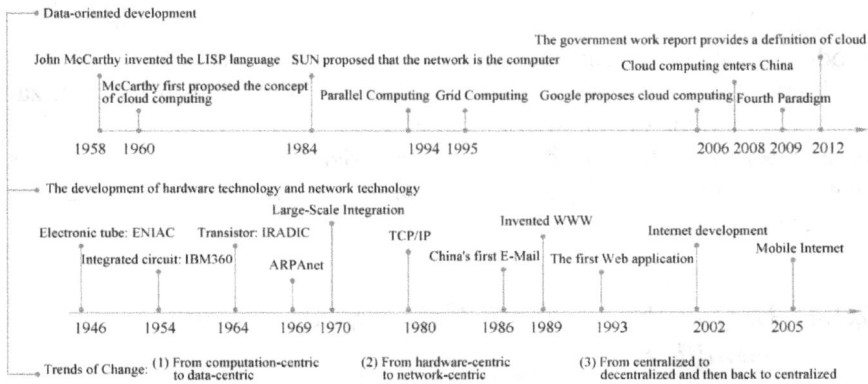

Fig. 2.3. The development journey of data-oriented technologies.

cross-sectional comparison of the evolution of hardware, networks, and cloud computing.

From the figure, it can be seen that in the early days of computer technology, due to the massive size and expensive prices of hardware devices, data generation was primarily the work of "individuals". In other words, data producers were mainly scientists or military departments, who were more concerned with the computational capabilities of computers. The level of computational capability determined the research capabilities and military strength of a country. During this period, the main driving force behind the development of computing technology was the evolution of hardware. This era witnessed a rapid progression of hardware from vacuum tubes to large-scale integrated circuits.

In 1969, the emergence of ARPANET (Advanced Research Projects Agency Network, developed by the U.S. Department of Defense, considered the precursor to the global Internet) changed the entire history of computer technology development. Gradually, the Internet became a significant force driving technological advancements, especially with the development and maturity of high-speed mobile communication networks, making data production a collective activity for the global human population. Anyone could generate and exchange data anytime, anywhere.

The core of network-based data composition became extremely complex, with diverse sources and a myriad of implicit correlations among different types of data. At this point, the data faced by computing became highly intricate. Various social applications began associating data with the complex operations of human society. As everyone became a data producer, social relationships and structures became implicit in the generated data. The production of data currently exhibits trends of popularization, automation, continuity, and complexity. The concept of big data emerged against this backdrop. A characteristic feature of this period is that computation must be oriented toward data, with data serving as the core element that structures the entire system.

2.1.4. *The fourth paradigm—Impact of big data on scientific research*

The emergence of the concept of big data has profoundly changed the mode of scientific research. In 2007, the late Turing Award winner Jim Gray, a pioneer in database theory, proposed the paradigm of data-intensive scientific research, known as "The Fourth Paradigm", as illustrated in Figure 2.4. He separated big data research from the third paradigm, which is computer simulation, and established it as an independent research paradigm. The reason for this separation is that the research methodology for big data differs from the traditional approach based on mathematical models.

The four paradigms of scientific research are illustrated in Figure 2.5. The first paradigm is experimentation, where knowledge is discovered through experiments, requiring minimal computation and data generation. The second paradigm is theory, where knowledge is discovered through theoretical research. For example, theories like Newtonian mechanics and Maxwell's electromagnetic field theory have allowed humanity to

Fig. 2.4.   Jim Gray, father of big data.

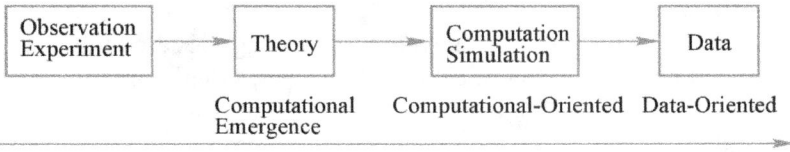

Fig. 2.5. The development history of the four paradigms of scientific research.

discover new planets not through observation but through computation, such as the discovery of Neptune and Pluto. The third paradigm is computation, where knowledge is discovered through computational methods. Using high-performance computing-based simulations, complex calculations like simulating nuclear explosions can be achieved. The fourth paradigm is data, where knowledge is discovered through data analysis. Utilizing massive datasets combined with high-speed computing enables the discovery of new knowledge, marking the era of data-intensive scientific discoveries.

Based on PB-scale big data, people can achieve discoveries without models and assumptions. By feeding these data into computer clusters with supercomputing capabilities, as long as the data are interconnected, statistical analysis algorithms can uncover new patterns, knowledge, and even laws that traditional scientific methods cannot discover. In fact, Google's advertising optimization configuration and AlphaGo, the artificial intelligence that defeated humans in the Go challenge in March 2016, have achieved this. This is the charm of the "Fourth Paradigm"! The transition from relying on human judgment to relying on data for decision-making reflects the impact of big data on scientific research and is one of the greatest contributions made by big data.

2.1.5. *Relationship between cloud computing and big data*

Cloud computing and big data are complementary concepts that describe two aspects of information technology in the data era: Cloud computing focuses on providing networked delivery methods for resources and applications; big data focuses on addressing the technical challenges brought by massive data volumes.

At the core of cloud computing is the business model, which essentially involves data processing technology. Data is an asset, and cloud computing provides the storage, access locations, and computing capabilities for data assets. Thus, cloud computing emphasizes the storage and computation of big data, as well as providing cloud computing services and running cloud applications. However, cloud computing lacks the ability to activate data assets, to extract value from data, and to perform predictive analysis on data, which are the core functions of big data. Cloud computing serves as the infrastructure architecture, while big data serves as the conceptual approach. Big data technologies will help people analyze and extract information from large, highly complex datasets, thereby discovering value and predicting trends for national governance, business decisions, and even personal life.

2.2. Characteristics of Big Data

The evolution of big data from structured data to semi-structured and unstructured data necessitates an analysis of its data characteristics to ensure data availability. The significant features of big data include large volume, high generation speed, complex data types, and low value density. Only data possessing these characteristics can be considered big data. The four "V" characteristics of big data are shown in Figure 2.6.

(1) *Volume*: Big data requires the collection, processing, and transmission of large amounts of data. Processing data in the PB range is quite common. The main sources of big data include internal business transaction information, commodity and logistics information in the online world, and human interaction and location information.

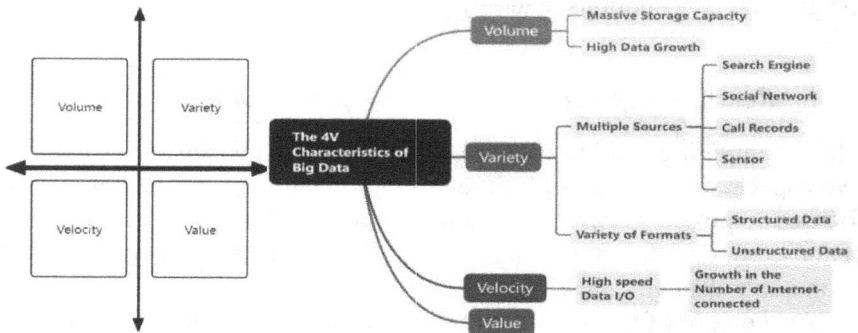

Fig. 2.6.   The 4V features of big data.

(2) *Variety*: Big data exhibits a wide variety of types and high complexity. It includes structured relational data, semi-structured web data, and unstructured video and audio data. Moreover, unstructured data is extensively present in social networks, the IoT, and e-commerce, with growth rates tens of times faster than structured data.

(3) *Velocity*: Big data needs to be collected, processed, and outputted frequently due to its temporal nature. Rapid processing is required to obtain results because data is time-sensitive. For example, if e-commerce data from a given day is not processed promptly, it could impact numerous immediate business decisions. Achieving real-time access to necessary information is crucial, with one second being the critical point. For many real-time big data applications, data must be processed within one second, otherwise the processing results become outdated and ineffective.

(4) *Value*: Big data has low value without appropriate processing. Extracting value from big data is akin to panning for gold in the sand. For instance, in the case of video data, only one or two seconds of useful data may be present within an hour of surveillance footage. Finding efficient algorithms to expedite the refinement of data value is a significant focus of current big data technology research.

2.3. Major Applications and Industry Driving Forces of Big Data

The primary objective of big data research is to utilize efficient information technology tools and computational methods to acquire, process, and analyze large datasets across various industries. This involves discovering and extracting the deep value inherent in the data, thereby providing industries with high-value-added applications and services.

2.3.1. *Primary applications of big data*

Big data holds vast prospects for applications across various industries, particularly in the public service sector. Implementing precision marketing through user behavior analysis is a typical application of big data.

(1) Internet enterprises can utilize big data technology to monitor and analyze daily generated data, amounting to hundreds of gigabytes, on

user clicks for online advertisements. This enables them to discern which users click on advertisements during which time periods, thereby assessing the value of ad placements and making timely adjustments.

(2) Smart grids can monitor users' electricity consumption data using big data technology. Smart meters collect and transmit these data to backend clusters every few minutes, where they are analyzed. This analysis yields insights into users' approximate electricity usage patterns, facilitating adjustments in electricity production to effectively prevent wastage of energy resources.

(3) Big data technology is applied in the field of connected vehicles. Vehicle terminals upload road condition data to backend data clusters every few minutes. Through analyzing these data, the system can determine the general road conditions and then push valuable road condition information to clients, helping them save time on the road.

(4) In the medical industry, every patient's medical history is recorded. By aggregating millions of cases nationwide and conducting data analysis and processing, patterns and rules can be identified. These patterns and rules are extremely helpful for doctors in diagnosing and treating various diseases.

2.3.2. *Enterprise-driven development of the big data industry*

The pioneers in the big data industry are the cornerstone of building the big data industry. Multinational giants such as IBM, Oracle, Microsoft, Google, Amazon, and Facebook are the main drivers of big data technology development.

(1) *Google*: Google provides all its software to users online. While users are using these products, their personal behaviors, preferences, and other information are collected by Google. Through big data technology, Google analyzes these data to better understand user needs. Therefore, the more diverse Google's product line becomes, the deeper its understanding of users, leading to more precise advertising placement and higher advertising value. Google's business model, which offers users easy-to-use, free software products in exchange for

understanding users and then generates revenue through precise advertising placement, disrupts the traditional model of selling software rights pioneered by Microsoft, making Google a giant in the Internet era.

(2) *IBM*: IBM's big data services include the following:
- data analysis, including text analysis;
- business event processing;
- monitoring of IBM Mashup Center;
- commercial services, such as Multimedia Mail Messaging (MMMS).

Among IBM's big data product portfolio is the series product InfoSphere BigInsights, based on Apache Hadoop and specifically designed for big data analysis. One software within this portfolio, called BigSheets, aims to help customers easily, simply, and intuitively extract and annotate relevant information from large amounts of data, tailored for industries such as finance, risk management, media, and entertainment.

(3) *Microsoft*: Microsoft's big data product concept is to "provide standardized products so that everyone can use data anytime, anywhere, and make better decisions". Microsoft believes that enabling everyone to gain insights from big data is crucial. To achieve this, Microsoft has introduced three major strategies for big data solutions: Big Control, Big Intelligence Pool, and Big Insight. Unlike other companies' approaches to handling big data, Microsoft advocates for thinking about the use of big data through the three methods of discovering data, analyzing data, and visualizing data.

(4) *Oracle*: Oracle's big data layout mainly consists of two aspects:
- providing end-to-end big data solutions from backend Hadoop and NoSQL to frontend data presentation for websites;
- combining traditional skills with new technologies to utilize Big Data SQL to offer SQL-on-Hadoop tools.

(5) *EMC*: EMC is a big data technology service provider listed on the New York Stock Exchange and Nasdaq. EMC's big data solutions encompass over 40 products.

(6) *Alibaba*: Alibaba possesses a vast amount of transaction and credit data. Its work in big data technology primarily focuses on building the underlying architecture for data circulation, collection, and sharing.

(7) *Huawei*: Huawei integrates high-performance computing and storage capabilities to provide a stable IT infrastructure platform for the mining and analysis of big data.

2.3.3. *Promotion of the big data industry by the Chinese government*

The era of big data has arrived. Facing the wave of business revolution driven by big data, one must either learn to utilize big data to create commercial value or risk being swept aside by the new business models driven by big data. The scale of data ownership and the ability to utilize data will become important components of national comprehensive strength, and control over data will also become a new focus of competition among countries and enterprises.

In December 2011, the Ministry of Industry and Information Technology of China released the "Twelfth Five-Year Plan for the Internet of Things", which proposed information processing technology as one of the four key technological innovation projects, including massive data storage, data mining, and intelligent analysis of images and videos, which are also important components of big data. In 2015, the State Council officially issued the "Action Plan for Promoting the Development of Big Data", which explicitly stated the promotion of big data development and application. It aims to create a new model of precise governance and multi-party cooperation in social governance over the next 5–10 years, to establish a new mechanism for smooth operation, to achieve safety and efficiency in economic operation, to build a new people-oriented system benefiting the entire population by offering livelihood services, to open up a new pattern of innovation driven by mass entrepreneurship, and to cultivate a new ecology of high-end intelligence and emerging prosperity in industrial development. This marks the formal elevation of big data to the national strategic level.

In 2016, the "Thirteenth Five-Year Plan for Big Data" was introduced. The plan solicited opinions from numerous experts and underwent thorough discussions and revisions. The contents of the plan include promoting the application of big data in the various links of industrial research and development, manufacturing, and the industrial chain, and supporting the service industry to utilize big data in establishing brands, precise marketing, and customized services.

The 19th National Congress of the Communist Party of China proposed "promoting the deep integration of the Internet, big data, artificial intelligence, and the real economy", which put forward higher requirements for the implementation of China's national big data strategy.

In August 2020, the China Electronics Information Industry Development Research Institute (CCID Group), directly under the Ministry of Industry and Information Technology, released the "White Paper on the Assessment of China's Regional Big Data Development Level (2020)", which focuses on three key areas of big data development: infrastructure, industrial development, and industry applications. It formed a comprehensive assessment index system of regional big data development in China, consisting of three primary indicators, 13 secondary indicators, and more than 30 tertiary indicators.

According to the data released by the Forward Industry Research Institute's "Analysis Report on the Development Prospect and Investment Strategy Planning of the Big Data Industry", the scale of China's big data industry reached 470 billion yuan in 2017. According to the data released by the China Big Data Industry Alliance in the "2021 China Big Data Industry Development Map and China Big Data Industry Development White Paper", in 2020, the scale of China's big data industry reached 638.8 billion yuan. It is estimated that the industry will maintain an average annual growth rate of over 15% in the next three years, and the industry scale will exceed 1 trillion yuan by 2023.

2.4. Key Technologies of Big Data

Big data technology is used to extract value from a large volume and variety of data through rapid collection, discovery, and analysis under economically affordable conditions. It represents a new generation of technology and architecture in the IT field. From the illustration of the big data industry structure (as shown in Figure 2.7), it can be seen that the processing of big data mainly includes data generation (also known as data collection or acquisition), data storage, data processing, and data application (also known as data analysis and mining). To accomplish these four tasks, computer support is required from hardware to software, with each layer performing different functions, necessitating corresponding technical support.

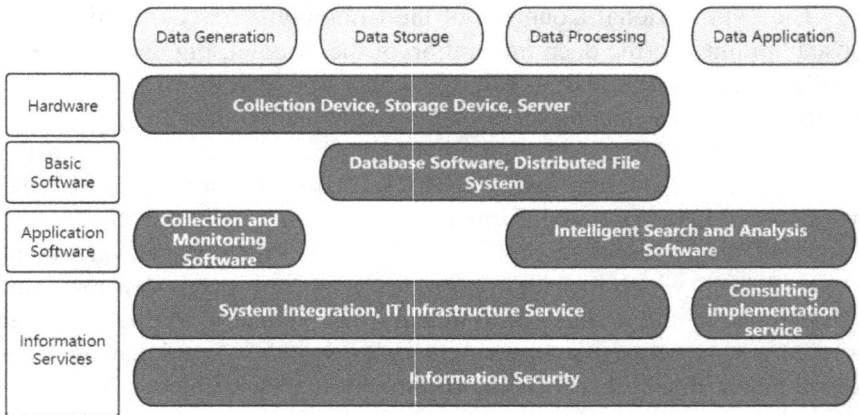

Fig. 2.7.   Schematic diagram of the big data industry structure.

The key technologies of big data mainly include the following five aspects:

(1) *Big Data Preprocessing Technology*: Big data preprocessing technology includes several categories:
   - *Data Collection*: Extract-Transform-Load (ETL) is a process that uses devices such as cameras and microphones to collect data from external systems and input them into an internal system interface. In today's rapidly evolving Internet industry, data collection has been widely applied in Internet and distributed fields.
   - *Data Access*: Relational databases, NoSQL, SQL, etc.
   - *Infrastructure Support*: Cloud storage, distributed file systems, etc.
   - *Presentation of Computational Results*: Cloud computing, tag clouds, relationship graphs, etc.
(2) *Big Data Storage Technology*: During the application process, data storage technology primarily utilizes a data stream formed by temporary files during processing. By searching for basic information and following a certain format, data records are stored on both external and internal computer storage media. Data storage technology requires naming based on relevant information characteristics and reflects the flowing data in the system in the form of a data stream,

synchronously presenting both static and dynamic data characteristics. Big data storage technology must also meet the following three requirements: storage infrastructure capable of persistent and reliable data storage; scalable access interfaces for users to query and analyze massive data; provision for efficient operations, such as querying, statistics, and updating, on both structured and unstructured massive data.

(3) *Big Data Analytics Technology*: Big data has a complex structure, with a majority of its composition being unstructured data. Relying solely on database Business Intelligence (BI) for analyzing structured data is no longer sufficient. Hence, there is a need for technological innovation, giving rise to big data analytics technology.

- *Data Processing*: This encompasses natural language processing techniques, multimedia content recognition technology, image-to-text conversion technology, geographic information technology, and others.

- *Statistics and Analysis*: Techniques include A/B testing, top N rankings, geographical distribution analysis, text sentiment analysis, semantic analysis, and more.

- *Data Mining*: This involves analyzing association rules, classification, clustering, and other methods.

- *Model Prediction*: Techniques such as predictive modeling, machine learning, modeling simulation, and pattern recognition.

(4) *Big Data Computing Technology*: Currently, over 85% of collected big data are unstructured or semi-structured data, which traditional relational databases cannot efficiently handle. Effectively processing unstructured and semi-structured data is a key point in big data computing technology. Another core issue for big data computing technology is how to perform cross-data type calculations.

Big data computing technology can be divided into batch processing and stream processing. Batch processing primarily operates on large, static datasets, returning results after the computation process is completed. It is suitable for tasks that require processing of all data before completion. Stream processing, on the other hand, computes data as they enter, without needing to operate on the entire dataset. It processes each data item as it is transmitted, making the results immediately available, and continues to update them as new data arrive.

2.5. Typical Big Data Computing Architectures

Currently, there are three typical big data computing architectures: Hadoop, Spark, and Storm.

Hadoop is an open-source computing framework under the Apache Software Foundation. Its strength lies in its ability to handle large-scale distributed data. All data to be processed are required to be local, meaning Hadoop's data processing operates at the disk level. The task processing involves high latency, indicating that Hadoop does not have an advantage in real-time data processing. Hadoop serves as the fundamental distributed computing architecture.

Storm is a real-time computing framework based on topology, processing data as they arrive—completely real-time and handling one piece of data at a time. Different mechanisms determine the suitability of Spark and Storm for different scenarios. For instance, in stock trading, where stock price changes are calculated not in seconds but in milliseconds, Spark has a real-time computation latency in seconds, making it unsuitable for such scenarios. On the other hand, Storm has a real-time computation latency in milliseconds, making it suitable for high-frequency stock trading scenarios.

Spark is an in-memory big data computing framework that enhances the real-time processing of data in a big data environment. It ensures high fault tolerance and scalability while processing data in near real-time. Spark collects data for a period and then processes it uniformly.

The aforementioned three typical big data computing architectures will be thoroughly explored in the subsequent chapters of this book.

Exercises

(1) Explain what are unstructured and semi-structured data.
(2) What are the three main components of the big data value chain?
(3) What are the 4Vs characteristics of big data?
(4) Briefly describe the relationship between cloud computing and big data.

Chapter 3

Virtualization

In recent years, there has been significant development and progress in the performance of computer hardware and software. The development of computer hardware has provided people with extremely powerful computing capabilities and abundant computing resources. If not effectively utilized, it can lead to resource waste. Meanwhile, with the advancement of computer software, there are increasing user scenarios where computers are used. This has resulted in users having more demands and requirements for computers, such as network security, data backup, system migration, system upgrades, and software and hardware costs, all of which need to be considered and resolved in the process of using computers. The emergence and application of virtualization technology provide users with perfect solutions to these problems. This chapter introduces virtualization technology, which is closely related to cloud computing and big data technology.

3.1. Introduction to Virtualization

Virtualization is not a recently introduced concept; in fact, virtualization technology has been around for a long time. It first appeared in the 1960s, when large computers already supported the simultaneous operation of multiple operating systems, each independently. Today, virtualization technology is no longer limited to merely supporting the simultaneous operation of multiple operating systems. It can help users save costs, improve software and hardware development efficiency, and provide greater convenience for users. Especially in recent years, virtualization technology has been widely applied in the fields of cloud computing and big data. There are many classifications of virtualization technology, and its different branches have emerged to cater to different user needs, such as network virtualization, server virtualization, and operating system virtualization. These different virtualization technologies effectively address users' practical needs.

3.1.1. *Concepts*

Virtualization encompasses various interpretations across industries and individuals. In the realm of computer science, it pertains to the abstraction of computer resources. Specifically, virtualization involves transforming a single physical computer into multiple logical machines through the application of virtualization technology. This enables the coexistence of multiple logical computers on a single physical device, with each capable of independently operating distinct operating systems and applications within their dedicated spaces. As a result, the efficiency of computational processes is significantly enhanced. Essentially, virtualization technology emulates tangible computer resources such as the CPU, memory, storage, network, and other discernible hardware components. When users utilize these resources facilitated by virtualization technology, the main distinction lies in the absence of physical interaction, as the functionality remains akin to its physical counterpart.

Virtualization facilitates the concurrent utilization of high-capacity, high-load, or high-traffic devices among multiple users. Each user is allocated an independent portion of resources that remain unaffected by others' activities. Although the data may be stored on the same physical device, the resources employed by each user are virtual and operate in isolation. For instance, in the context of a virtual hard disk, users leverage a virtualized disk provided by the virtualization technology. From the users' perspective, this disk functions as a genuine and usable storage medium. While these virtual disks could be stored as separated files within physical storage, each user can solely access their own disk and is unable to infringe upon others', ensuring data safety and independence. Notably, there are distinctions in the network interfaces, network resources, and operating systems employed by each user.

Virtualization technology enables the consolidation of numerous scattered resources into a unified entity, making users perceive them as an integrated whole. A prime example is storage virtualization, wherein multiple physical hard drives are amalgamated and presented to users as a single, comprehensive virtual hard drive.

The use of virtualization technology enables the dynamic allocation of resources, allowing for the dynamic scaling up or down of resources for a particular user. If a user has a need, such as requiring additional disk space or more network bandwidth, virtualization technology can quickly meet the user's demand by adjusting the corresponding configurations without interrupting the user's business operations.

As virtualization technology is applied in various systems and environments, its advantages in both commercial and scientific aspects are becoming increasingly evident. Virtualization technology reduces operational costs for enterprises while enhancing system security and reliability. It enables enterprises to serve users with greater flexibility, speed, and convenience, and users themselves are more willing to embrace the variety of benefits brought about by virtualization technology. To provide a more intuitive understanding of virtualization, a simple comparison between a computer system with and without virtualization technology is illustrated in Figure 3.1.

As shown in Figure 3.1(a), without the application of virtualization technology, the operating system is directly installed on the hardware, and the application programs run within the operating system. The application programs monopolize the entire hardware platform. On the other hand, when employing virtualization technology, an additional layer of virtualization middleware is introduced, which provides hardware simulation. In this way, the installation of multiple operating systems and applications

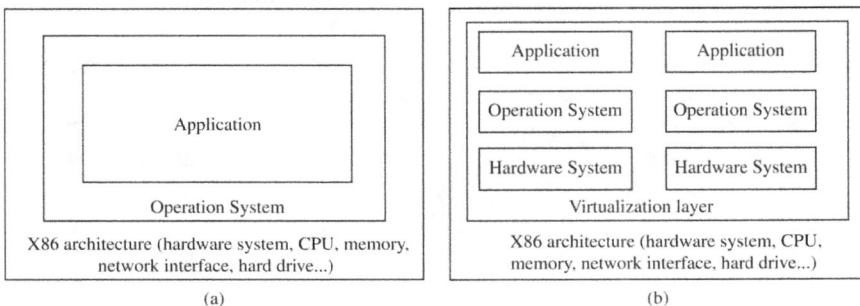

Fig. 3.1. Comparison of virtualization software and hardware frameworks: (a) Without applying virtualization technology and (b) applying virtualization technology.

on this virtualization layer is allowed, with each of them operating independently, as depicted in Figure 3.1(b).

Virtualization technology can simulate multiple different hardware systems at the same time, while the operating system is installed on the virtualized hardware systems. This prevents the operating system and applications from exclusively occupying the entire hardware resources, enabling multiple operating systems to run simultaneously.

Table 3.1 presents a comparison of a computer system prior to and following the implementation of virtualization technology.

3.1.2. Classification of virtualization technology

Virtualization is an abstract concept that can have different solutions tailored to different industries and needs. The most popular virtualization technologies now include server virtualization, network virtualization, storage virtualization, and operating system virtualization.

Virtualization technologies have formed their own hierarchical structure based on various purposes, as shown in Figure 3.2. From bottom to top, they include network virtualization, storage virtualization, server

Table 3.1.   Comparison before and after the application of virtualization technology.

Before using virtualization technology	After using virtualization technology
Only one operating system can run on a computer at a time	The operating system, applications, and hardware systems are mutually independent
Software is tightly coupled with hardware	Software and hardware are loosely coupled
All applications running within the same operating system can often cause compatibility issues, especially with security applications	A computer can run multiple operating systems, with each operating system running independently and not affecting each other. These operating systems can be the same or different
Low hardware resource utilization, such as CPU usage often being at 10% or lower	It has increased the utilization of hardware resources and saved a lot of costs
Not easy to scale and maintain, with high maintenance costs	Easy to expand and maintain, with low maintenance costs
The program runs faster	The program runs a bit slowly

User Experience Virtualization
Application Virtualization
Desktop Virtualization
Service Virtualization
Operating System Virtualization
Server Virtualization
Storage Virtualization
Network Virtualization

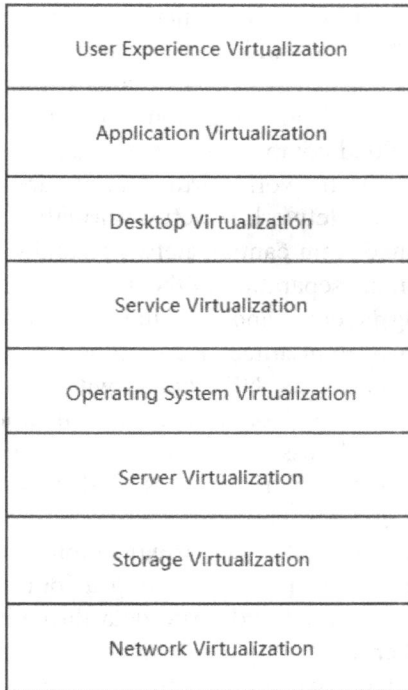

Fig. 3.2. Classification of virtualization technologies.

virtualization, operating system virtualization, service virtualization, desktop virtualization, application virtualization, and user experience virtualization. Each layer corresponds to specific use cases, and users can choose different virtualization technologies based on their needs. This hierarchical structure is essentially a classification of virtualization technologies.

(1) *Network Virtualization*: Network virtualization integrates network resources; in simple terms, it consolidates hardware and software network device resources, as well as network functions, into a unified, software-managed virtual network. Network virtualization is a type of computer network that includes at least partially virtual network connections. Virtual network connections refer to network connections between multiple computing devices that do not involve physical connections but are rather achieved through network virtualization. There

are two common types of virtual networks: protocol-based (such as VXLAN, VLAN, VPN, and VPLS) virtual networks and virtual device-based (such as network connections within a Hypervisor) virtual networks. Network virtualization is commonly applied in large servers, such as cloud computing servers. Currently, the most mature overall solutions in network virtualization are Software Defined Network (SDN) and Network Function Virtualization (NFV).

SDN originated from campus networks and matured in data centers. It focuses on the separation of the network control plane and data plane, addressing layers 2 and 3 of the OSI model. SDN optimizes network infrastructure architecture, such as Ethernet switches, routers, and wireless networks. NFV began with large service providers and focuses on the virtualization and generalization of network forwarding functions, addressing layers 4 and 7 of the OSI model. NFV optimizes network functions, such as load balancing, firewall, and WAN optimization controllers.

(2) *Storage Virtualization*: Storage virtualization integrates all storage resources into a storage pool, providing a logical storage interface externally. Users can read and write data through this logical interface, and no matter how many hardware storage devices are involved, users only see one unified storage resource from the outside perspective. The simplest understanding of storage virtualization is to abstract the storage hardware resources by integrating one or more target services or functions with additional functionalities, providing comprehensive service functionality in a unified manner.

For users, virtualized storage resources are like a huge "storage pool". Users will not see specific disks, do not know how many disks there are, and do not need to care about where their data are specifically stored on which disk. A schematic diagram of storage virtualization is shown in Figure 3.3.

After storage virtualization, users see only one hard disk, and they only need to interact with this single hard disk. As for where the user's final data are stored, that is determined by the storage virtualization management program.

There are mainly two ways to implement storage virtualization: block virtualization and file virtualization. Block virtualization mounts remote disk blocks locally through Storage Area Networks (SANs), such as an Internet Small Computer System Interface (ISCSI). Then, through a Logical Volume Manager (LVM), these disk

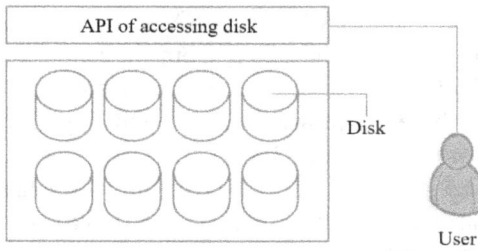

Fig. 3.3. Storage virtualization.

blocks are combined to form a new disk. File virtualization mounts remote file system paths locally through SANs, such as Network File Systems (NFSs) and Service Message Blocks (SMBs). Locally, what is seen are files under the specified paths, not a disk block.

(3) *Server Virtualization*: Server virtualization, sometimes also known as platform virtualization, abstracts server physical resources into logical resources, allowing one server to become several or even hundreds of isolated virtual servers. Users are no longer constrained by physical boundaries, enabling CPU, memory, disk, I/O, and other hardware to become a dynamically managed "resource pool", thereby improving resource utilization, simplifying system management, achieving server consolidation, and making IT more adaptable to changes in business.

Server virtualization actually involves packaging the operating system and applications into virtual machines (VMs), enabling these components to have good mobility. A VM refers to a complete computer system running in a fully isolated environment with full hardware functionality simulated by software. The operating system running within a VM is called the guest operating system (Guest OS), while the platform managing these VMs is known as the VM Monitor (VMM), also referred to as the Hypervisor. The operating system running the VMM is called the host operating system (Host OS). The VMM is the core of VM technology, as it is a layer of code situated between the operating system and computer hardware, responsible for partitioning the hardware platform into multiple VMs. The VMM not only manages the running state of VMs but also allows for customization of parameters such as CPU count and memory size for each VM.

Not all server virtualization requires a Host OS. Some virtualization products can run directly on bare metal, meaning they run on top of the hardware, such as VMware ESX series, Xen Server, and other virtualization products. Therefore, from the perspective of whether a Host OS is included, server virtualization can be divided into two types: Type 1 (bare-metal architecture) and Type 2 (hosted architecture), as illustrated in Figure 3.4.

- *Bare-Metal Architecture*: This virtualization layer runs directly on the hardware, that is, bare metal. It is also known as bare-metal framework. The hypervisor is installed directly on the bare metal, which can significantly improve running efficiency. Users can manage the virtualization layer through a console or a web page. For users, each user will feel like they are operating on an independent computer isolated from other users, even though in reality, the service for each user is provided by the same machine. In this case, a VM is an operating system managed by a potential control program. Examples include VMware ESX and Xen.
- *Hosted Architecture*: In the hosted architecture, the virtualization layer runs within an operating system (Host OS). The VM needs to be installed into the operating system as an application and run as an application. Therefore, it is visible as an application running processes within the operating system. The operation of the VM depends on the current physical resources of the host and the virtualization support, and its efficiency is lower than that of the bare-metal architecture.

From the perspective of the number of servers and virtual applications, server virtualization can be divided into three types. The first type refers to one server being virtualized as multiple

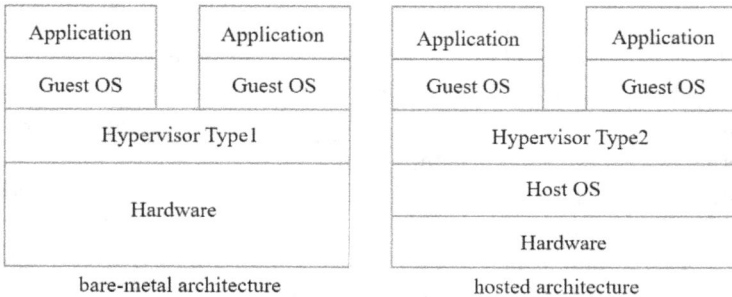

Application	Application	Application	Application
Guest OS	Guest OS	Guest OS	Guest OS
Hypervisor Type1		Hypervisor Type2	
Hardware		Host OS	
		Hardware	
bare-metal architecture		hosted architecture	

Fig. 3.4. Two types of server virtualization.

servers, dividing a physical server into multiple independent and mutually exclusive virtual environments. The second involves multiple independent physical servers being virtualized as one logical server, enabling multiple servers to collaborate in processing the same business. Additionally, the third entails multiple physical servers being virtualized as one logical server, which is then divided into multiple virtual environments where multiple businesses run on multiple virtual servers.

From the degree of virtualization, server virtualization can be divided into full virtualization, paravirtualization, and hardware-assisted virtualization.

- *Full Virtualization*: It refers to the VM simulating the entire underlying hardware, including the processor, physical memory, network card, and display, and completely simulating a real computer hardware device. This allows an operating system running within the VM to operate normally without any modifications, as if it were running in a real physical environment.

- *Paravirtualization*: The emergence of paravirtualization was due to the fact that full virtualization could not meet the requirements when executing certain privileged operations due to the long execution time. In order to reduce the execution time of the guest machine, paravirtualization allows the guest machine to run certain time-consuming instructions directly on the real host machine or real hardware, thus improving execution efficiency and reducing instruction execution time. To achieve paravirtualization, some modifications need to be made to the guest machine's operating system so that the guest machine can run instructions directly on real hardware or the host machine when running by using the paravirtualization API provided by the VMM.

- *Hardware-Assisted Virtualization*: It is a feature provided for the CPU, specifically designed to enhance the efficiency of VM operations, allowing VMs to execute privileged instructions more quickly and reducing excessive context switches and emulation. Common hardware-assisted virtualization technologies include Intel VT and AMD-V. They provide specific instructions that the VMM can leverage to improve the efficiency of VM operations. Nowadays, on the x86 platform, many virtualization technologies utilize this assistance feature, such as KVM, VMware, and Xen, that make use of these special instructions.

(4) *Operating System Virtualization*: Operating system virtualization refers to running one or more independent users simultaneously on the same operating system, each with their own runtime environment. Each user can only run applications within their own permission scope, and they are not affected by each other. Users can access their own resources through remote desktop, but they share the same operating system. An illustration of operating system virtualization is shown in Figure 3.5. Typical examples include Docker, Windows Server 2008, and Ubuntu Server.

(5) *Service Virtualization*: Service virtualization is a virtual application that is hardware-independent and is implemented for software. Service virtualization is hidden from end users and essentially provides services through virtualization, such as firewalls, load balancing, databases, and data storage. For instance, firewall service virtualization is known as FireWall as a Service (FWaaS), and load balancing virtualization is known as Load Balance as a Service (LBaaS).

(6) *Desktop Virtualization*: Desktop virtualization refers to virtualizing the terminal system of a computer (also known as a desktop) to achieve security and flexibility in desktop usage. Personal desktop systems can be accessed over the network from anywhere and at any time using various devices such as PCs, tablets, and smartphones.

Users can access several different desktop systems at the same time on the same physical device, which can have the same or different operating systems. The server will be independent of the users' desktops, with each user having their own user space and not affecting each other. Independent desktops and the corresponding application software with the users can realize remote access to the desktop. Virtual desktop access typically relies on remote connections or thin-client devices.

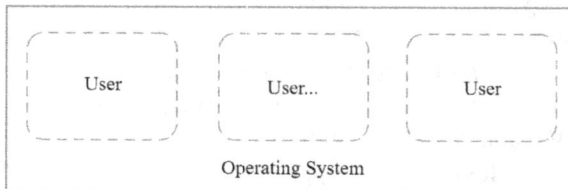

Fig. 3.5.   Illustration of operating system virtualization.

(7) *Application Virtualization*: Application virtualization means that the same application can run normally on different CPU architectures and different operating systems. Application virtualization is also a software technology, which is an encapsulation that is independent of the underlying operating system. Application virtualization enables cross-platform compatibility, allowing a single codebase to run seamlessly across different operating systems and CPU architectures. A key example is the Java Virtual Machine (JVM), which allows Java applications to be developed and executed solely within a JVM environment, eliminating platform-specific dependencies.

Similarly, there are many other software that support this type of application virtualization, such as Python and Wine (which is more common in Linux; if you want to run Windows applications in Linux under the circumstance without installing a VM, you can install a Wine environment and then run Windows applications directly).

(8) *User Experience Virtualization*: User experience virtualization, sometimes also called user virtualization, refers to the same content or interface that users see on different devices such as laptops, tablets, or phones. Modifications made on one device will result in consistent configurations or modifications on another device. Users' relevant information and application configurations will be synchronized to the respective user devices.

3.1.3. *Advantages and disadvantages of virtualization technology*

(1) *Advantages*: The advantages of virtualization technology are mainly reflected in the following aspects:

- Reduce physical resource investment and save costs: When users need different operating systems or more computing resources, they can simply add several different operating systems through VMM. When these systems are no longer needed, they can be directly shut down or deleted.
- Convenient migration of virtual data resources: Virtual data resources (typically data generated by VMs) can be easily migrated to other data centers without affecting the virtual data resources themselves. If there is data migration or equipment damage involved, there is no need to physically migrate equipment. Simply backup the data generated by the VMM to another

data center, and then manage it through VMM. The previous data and services will not be affected in this way.

- Increase the utilization rate of physical resources: The average CPU utilization of traditional server hosts is below 10%, and the CPU usage is far from reaching an ideal state, which results in significant resource waste. By using virtualization technology, multiple servers can be deployed on the same physical device, thereby increasing the utilization of this physical device and significantly reducing cost expenses.

- More environmentally friendly and saves energy: By applying virtualization technology, it is possible to reduce the investment in physical hardware, thereby lowering the electricity consumption and physical space occupied by the hardware and making it more environmentally friendly.

- Easy to automate maintenance and operation, reducing maintenance costs: Virtualization technology simulates physical devices through software. Virtual resources implemented in software can be automatically maintained and managed through corresponding interfaces. This can improve work efficiency and reduce maintenance costs.

- Data security is more guaranteed: Each virtualized device will have corresponding files generated in the physical device. Administrators only need to back up the data accordingly and manage it regularly to ensure the security of the data. In the event of uncontrollable situations such as natural disasters damaging the physical equipment, administrators only need to restore the already backed-up data to a new device to recover the user's data. If the architecture is designed properly, it may not even interrupt user work to update and replace the system.

However, virtualization technology is not omnipotent. It also has problems that it cannot solve and is not suitable for all users. Some of the flaws in the current development of virtualization technology are summarized in the following.

(2) *Disadvantages*:
- Currently, there is no unified virtualization technology standard or platform, and there are no open protocols either. Various

virtualization technology providers exist in the market, such as Microsoft, VMware, and Xen. The operational effectiveness of the virtualization technologies they use varies, and they are mutually incompatible.

- If data are not backed up, there will be certain risks in applying virtualization technology. Although this technology can achieve data backup, it is ultimately based on real hardware systems. If multiple applications and servers are placed on the same physical device and that device encounters issues without redundant physical devices as backups, all applications and services will become unusable.
- The migration of a virtual data center, especially that of online services, has a significant impact on users. Due to its large data volume, numerous applications, and complex structure, once migrated, it may lead to many unforeseeable consequences.

3.1.4. *Virtualization technology and cloud computing*

Virtualization technology is an important supporting technology for cloud computing. Cloud computing is a model for the increased use and delivery of Internet-based services. In cloud computing, dynamic and scalable virtualized resources are provided through the Internet. Using virtualization technology, applications and data can be presented to users in different ways at different levels, offering convenience for users and developers of cloud computing.

The main function of virtualization is to abstract a single resource into multiple ones for users to use, while cloud computing helps different departments (through private clouds) or companies (through public clouds) access an automatically provisioned resource pool. With virtualization technology, users can create multiple simulated environments or dedicated resources based on a single physical hardware system. Cloud computing is a combination of various rules and methods that can provide users with computing, networking, and storage infrastructure resources, services, platforms, and applications on demand across any network, which come from the Internet. In simple terms, cloud computing can be seen as a virtual resource pool managed by a series of automated software, designed to help users access these resources as needed through self-service that supports automatic scaling and dynamic resource allocation.

Cloud computing provides services, and virtualization technology is the technical support for cloud computing. In the deployment plan of cloud computing, virtualization technology can make IT resources apply more flexibly. Meanwhile, during the application process of virtualization technology, cloud computing also provides resources and services on demand. In some specific scenarios, cloud computing and virtualization technology cannot be separated; they need to be combined to better meet customer needs. Through virtualization technology, cloud computing transforms computing, storage, applications, and services into resources that can be dynamically configured and scaled, thus presenting them in a logically unified service form to users. Therefore, virtualization technology is the extremely important and core driving force behind cloud computing.

3.2. Principles of Virtualization Technology

So far, virtualization technology has made progress in various aspects. Virtualization has evolved from purely software virtualization to processor-level virtualization, then to platform-level virtualization, and even to input/output virtualization. For data centers, virtualization can save costs, maximize the utilization of data center capacity, and better protect data. Virtualization technology has become the foundation of private and hybrid cloud design solutions.

This section briefly introduces the principles of virtualization technology, including the principles of VMs, CPU virtualization, memory virtualization, and network virtualization.

3.2.1. *Principles of VM technology*

VM refers to a complete computer system that simulates the functions of a hardware system through software, running in a fully isolated environment. In a nutshell, a VM is a computer simulated on the host machine through software. VM technology is a resource management technique that abstracts and transforms various physical resources of a computer, such as servers, networks, memory, and storage, breaking down the inseparable barriers between physical structures and allowing users to apply these resources in a better way. The new virtual parts of these resources are not limited by the existing setup, location, or physical configuration of resources. Generally, virtualized resources refer to computing power and data storage. In actual production environments, VM technology is mainly used to address the excess capacity of cloud data centers and high-performance physical hardware, as well as the reorganization and reutilization of old hardware with low capacity, transparentizing the underlying physical hardware in order to maximize the utilization of physical hardware. By integrating multiple operating systems into a high-performance server, one can maximize the utilization of all resources on the hardware platform, achieve more applications with less investment, simplify IT architecture, reduce the complexity of resource management, and avoid unnecessary expansion of IT architecture. Furthermore, the hardware-independent feature of VMs allows for real-time migration during VM operation, enabling true uninterrupted operation and maximizing business continuity without incurring the high cost of purchasing ultra-high-availability platforms.

VM technology enables a computer to run multiple operating systems simultaneously, with multiple programs running in each operating system, all running on a virtual CPU or virtual host. VM technology requires support from the CPU, motherboard chipset, BIOS, and software such as VMM software or certain operating systems themselves.

The core of VM technology is the VMM, also known as the Hypervisor. The role of the VMM is to allocate access to the host machine's hardware resources for lower users and manage the VM's operating system and applications for higher users. It is a host program, which is a layer of code located between the operating system and the computer hardware, used to divide the hardware platform into multiple VMs, enabling a single computer to support multiple identical execution environments. Each user will feel like they are operating on a separate and isolated computer, even though in reality they are all using the same machine. In this scenario, each VM is managed by a potential control program running the operating system. The VMM virtualizes a separate set of virtual hardware environments (including processor, memory, and I/O devices) independent of the actual hardware for each customer operating system. The VMM uses a scheduling algorithm to share the CPU among the VMs, such as using a time-slice round-robin scheduling algorithm.

The system is no different from an actual computer operating system and can also be infected by viruses. However, due to the fact that the VM is a closed virtual environment, if the VM is not connected to the host machine, it will not be affected by viruses from the host machine.

3.2.2. *Principles of CPU virtualization*

The primary purpose of CPU virtualization is to allow multiple VMs to run simultaneously in the VMM. CPU virtualization technology simulates a single CPU as multiple CPUs, enabling all VMs running on top of the VMM to operate simultaneously and independently of each other, without mutual interference, in order to enhance the efficiency of computer usage. In the computer system, the CPU is the core of the computer, and without it, the computer cannot function properly. Therefore, the key to whether a VM can run properly is whether the CPU can be successfully emulated.

From the perspective of design principles, the CPU mainly consists of three major parts: the arithmetic logic unit, the control unit, and the processor registers. Each type of CPU has its own Instruction Set Architecture (ISA), and each instruction executed by the CPU is based on the corresponding instruction standards provided by the ISA. The ISA mainly consists of two instruction sets: the user instruction set (User ISA) and the system instruction set (System ISA). The user instruction set generally refers to ordinary arithmetic instructions, while the system instruction set generally refers to instructions for handling system resources. Different instructions require different permissions, and the instruction's effectiveness is manifested when executed within the corresponding permission level. In the x86 architecture framework, CPU instruction permissions are generally divided into four levels: ring0, ring1, ring2, and ring3, as shown in Figure 3.6.

The most commonly used CPU instruction privilege levels range from 0 to 3: Instructions in privilege level 0 are typically only executable by the kernel, while instructions in privilege level 3 are for regular user

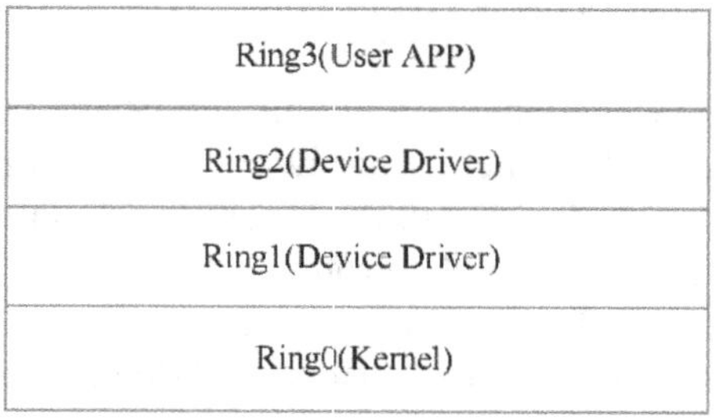

Ring3(User APP)
Ring2(Device Driver)
Ring1(Device Driver)
Ring0(Kernel)

Fig. 3.6.   Four instruction privileges of the CPU.

execution. Privilege levels 1 and 2 are generally utilized by device drivers. The following are three scenarios that can arise:

- asynchronous hardware interrupts, such as disk reads and writes;
- system calls, such as int and call;
- exceptions, such as page fault.

From the above discussion, it can be seen that in order to achieve CPU virtualization, the main problem is the permission issue of the system ISA. Ordinary ISAs do not need simulation; they only need to protect the CPU running state, ensuring the separation of state between each VM. ISAs that require permission need to be captured and simulated. Therefore, to achieve CPU virtualization, the following problems need to be addressed:

- All accesses to the ISA of the VM system need to be emulated by the VMM through software. That is, all instructions generated on the VM need to be emulated by the VMM.
- The system state of all VMs must be saved to memory through the VMM.
- All system instructions need corresponding functions or modules at the VMM to simulate them.
- The capture and simulation of CPU instructions are key to solving the CPU privilege problem, as shown in Figure 3.7.

When the CPU is executing instructions normally, if it is a regular instruction, it can be executed directly without simulation. However, when encountering instructions that require privileges, they will be caught by the VMM, and control will be transferred to the VMM, which will determine how to execute these privileged instructions. The VMM will

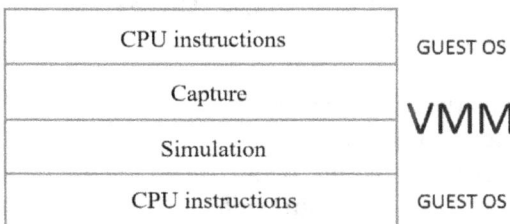

CPU instructions	GUEST OS
Capture	**VMM**
Simulation	
CPU instructions	GUEST OS

Fig. 3.7. Capture and simulation of CPU instructions.

generate a series of instruction sets related to the instruction and execute them during the simulation. After the VMM has finished execution, control is handed back to the guest operating system, as shown in the following example:

```
mov ebx, eax;   # normal instruction
cli;            # instruction that requires permissions
need to be captured and simulated
mov eax, ebx;   # normal instruction
```

When virtualized, it dynamically changes to a command similar to the following:

```
mov ebx, eax;      # normal instruction
call handle_cli;   # replace it with an instruction
from the VMM that, after handle_cli has finished
executing, continues to execute the instruction
afterwards
move eax, ebx;     # normal instruction
```

The real situation is not that simple, as not all CPU frameworks support similar capture, and the performance burden of capturing such privilege operations can be huge. Moreover, while virtualizing instructions, it is also necessary to implement the privilege level of the CPU in the physical environment, i.e., virtualize the execution privilege level of the CPU. Without the support of the privilege, the effect of the instruction execution may not be the desired effect.

In order to improve the efficiency and execution speed of virtualization, the VMM implements a Binary Translator (BT), which is responsible for the conversion of instructions, and a Translation Cache (TC), which is used to store the translated instructions. The BT can be used to convert instructions in the following ways:

(1) For normal instructions, they are copied directly into the TC, which is called an "ident" conversion.
(2) For instructions that require permissions, the conversion is done by replacing some of the instructions, which is called an "inline" conversion.
(3) For instructions that require permissions, users need to simulate them through an emulator and pass the results of the simulation to the VM

in order to achieve the effect of virtualization. This is called a "call-out" conversion.

Because the instructions need to be simulated, some operations consume a longer time, and in the case of full virtualization, it will be less effective in its execution, which is why there are two other virtualization techniques, semi-virtualization and hardware-assisted virtualization, used to improve the efficiency of VM operation.

3.2.3. *Principles of memory virtualization*

In addition to CPU virtualization, another key virtualization technology is memory virtualization. Memory virtualization allows each VM to share physical memory, which can be dynamically allocated and managed by the VMM, ensuring that each VM has its own separate memory operating space. Memory virtualization for VMs is somewhat similar to virtual memory management in operating systems. In the operating system, there is no connection between the memory address space "seen" and used by the application program and whether these addresses are contiguous in the physical memory or not. The operating system maintains virtual-to-physical address mappings in page tables. During memory allocation requests from applications, hardware components—specifically the Memory Management Unit (MMU)—perform address translation. The Translation Lookaside Buffer (TLB) accelerates this process by caching frequently accessed mappings. The MMU and TLB automatically translate the requested virtual address into the corresponding physical address. Currently, all x86-based CPUs include MMUs and TLBs to improve the efficiency of mapping virtual addresses to physical addresses. Therefore, memory virtualization should also address the MMU and TLB together in the virtualization process. Multiple VMs run on the same physical device, and there is only one real physical memory. At the same time, it is necessary to make each VM run independently; therefore, it is necessary for the VMM to provide virtualized physical addresses, that is, to add another layer of physical virtual addresses, as shown in Figure 3.8.

- *Virtual address*: The address used by the guest VM application.
- *Physical address*: The physical address provided by the VMM.
- *Machine address*: The real physical memory address.
- *Mapping relationship*: Consists of two parts, the mapping of virtual addresses in the client to VMM physical addresses and the mapping of VMM physical addresses to machine addresses.

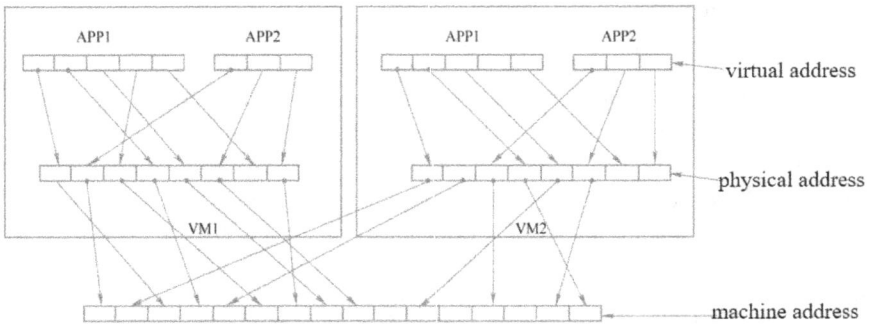

Fig. 3.8.   Memory allocation of VMs.

From Figure 3.8, it can be seen that the guest VM can no longer directly access the physical addresses of the machine through the MMU. The physical addresses it accesses are provided by the VMM. That is, the operations of the guest VM remain unchanged, maintaining the conversion from virtual addresses to physical addresses. However, before obtaining the real address, an additional address conversion is required (the VMM physical address to machine address conversion). By performing two memory address translations, independent operation of guest VMs can be achieved, although their efficiency will be significantly lower. In order to improve efficiency, shadow page tables are introduced, and later hardware-assisted virtualization further enhances the efficiency of address mapping queries. We will not delve further into this topic.

3.2.4. *Principles of network virtualization*

Network virtualization provides virtual network devices implemented in software, through which virtualization platforms can communicate with other network devices. The communication targets can be real physical network devices or virtual network devices. Therefore, network virtualization aims to establish virtual connections between devices that are independent of physical connections. The main issues addressed by network virtualization are network device virtualization and virtual connections. Virtualized network devices can be individual network interfaces, virtual switches, virtual routers, and so on. Within the same Local Area Network (LAN), any two different virtual devices can establish network connections. If they are not in the same network, network protocols such as Virtual Local Area Network (VLAN) and Virtual Private Network (VPN) are needed to achieve normal network connections and communication.

Using VLAN as an example to briefly explain the connection and communication of network virtualization. VLAN divides network nodes into several logical workgroups as needed, with each logical workgroup corresponding to a virtual network. Each virtual network functions like a LAN, where different virtual networks are independent of each other and cannot connect or communicate. If communication is required, routing devices are needed to assist in forwarding packets for proper communication. Since these groupings are all logical, the devices are not limited by physical location and only require support from network switching devices.

3.2.5. *Principles of CGroups*

(1) *What are CGroups*? CGroups represent a mechanism provided by the Linux kernel, which can integrate (or separate) a series of system tasks and their subtasks into different groups based on resource allocation levels according to needs, thereby providing a unified framework for system resource management. In simple terms, CGroups can limit and record the physical resources (including CPU, memory, and IO) used by task groups, providing assurance for virtualization in containers and serving as the foundation for building a series of virtualization management tools like Docker.

For developers, CGroups have the following four characteristics:
- The API of CGroups is implemented in a pseudo filesystem manner, allowing user-space programs to organize and manage CGroups through file operations.
- The organization and management operations of CGroups can be granular down to the thread level. Additionally, users can create and destroy CGroups, enabling resource reallocation and management.
- All resource management functions are implemented in a subsystem manner, with a unified interface.
- At the beginning of its creation, the subtask is in the same control group as the parent task.

Essentially, CGroups are a series of hooks attached to the kernel on top of programs, triggering corresponding hooks during program runtime to achieve the purpose of resource tracking and limitation.

(2) *The Role of CGroups*: The main purpose of implementing CGroups is to provide a unified interface for resource management at different user levels. From resource control of individual tasks to operating system-level virtualization, CGroups offer the following four major functions:

- *Resource Limitation*: CGroups can limit the total amount of resources used by tasks. For example, a maximum limit is set on the memory usage of an application at runtime, and once this quota is exceeded, an Out Of Memory (OOM) alert is triggered.

- *Priority Allocation*: This is achieved by assigning the number of CPU time slices and the size of disk IO bandwidth, which is, in fact, equivalent to controlling the priority of task execution.

- *Resource Statistics*: CGroups can track the system's resource usage, such as CPU usage time and memory usage, making it highly suitable for billing purposes.

- *Task Control*: CGroups can perform operations such as suspending and resuming task execution.

(3) *CGroups Terminology*:

- *Task*: In the terminology of CGroups, a task represents a process or thread.

- *Cgroup*: Resource control in CGroups is implemented on a per-CGroup basis. A CGroup represents a control group divided according to certain resource control standards, containing one or more subsystems. A task can join a CGroup or migrate from one CGroup to another.

- *Subsystem*: The subsystems in CGroups are resource scheduling controllers. For example, the CPU subsystem can control CPU time allocation, and the memory subsystem can limit memory usage within a CGroup.

- *Hierarchy*: The hierarchy is formed by a series of CGroups arranged in a tree structure, with each level controlling resources by binding to corresponding subsystems. CGroup nodes within a level can include zero or multiple child nodes, with child nodes inheriting the subsystems mounted by the parent node. The entire operating system can have multiple levels.

3.3. *Virtualization technology solutions*

With the development and application of virtualization technology, various virtualization technology solutions have emerged in the market. The following is an overview of these common virtualization technology solutions.

3.3.1. *OpenStack*

OpenStack (https://www.openstack.org/) is a free software and open-source project licensed under the Apache License, initiated and developed through a collaboration between the National Aeronautics and Space Administration (NASA) and Rackspace. Its logo is shown in Figure 3.9. OpenStack is an open-source cloud platform that manages VMs through corresponding APIs and drivers, supporting almost all types of virtualization environments on the market. OpenStack itself does not provide virtualization functions; these are provided by the VMM, with OpenStack managing the VMM through the respective APIs. OpenStack is responsible for building the platform and enhancing peripheral functions. The original intention behind the design of OpenStack was to adapt to the architecture of distributed applications, where components of applications can span across multiple physical or virtual devices within the platform. These types of applications are also designed to scale with the addition of application instances or the rebalancing of the load between application instances. The goal of OpenStack is to provide a cloud computing management platform that is simple to implement, scalable, feature-rich, and standardized.

Fig. 3.9 OpenStack logo.

From a logical perspective, OpenStack consists of three main components: the control module, the network module, and the compute module. The control module primarily runs API interface services, message queues, database management modules, and web interfaces; the network module provides network services for each VM; and the compute module is responsible for processing messages and controlling VM operations.

In terms of its composition, OpenStack includes numerous modules, all of which can be deployed in a distributed manner. Some of the key modules include Nova, Keystone, Ceilometer, Horizon, Glance, Neutron, Cinder, and Swift. Nova primarily provides compute functionality; Keystone handles authentication and authorization; Ceilometer monitors resources and system operations; Horizon offers a user-friendly web platform for management; Neutron is responsible for building network environments and virtualization; Glance manages image files; Cinder handles block storage and can provide Storage as a Service (SaaS) to users; Swift also manages storage, focusing on data objects, images, data backups, and other data storage used by the platform, and can also backup storage for Cinder data.

3.3.2. *KVM*

The Kernel-based VM (KVM) is open-source software, and its logo is shown in Figure 3.10. The official website address is https://www.linux-kvm.org/page/Main_Page. KVM is a Linux full virtualization solution based on the x86 architecture and hardware virtualization technology. Hardware virtualization technology is provided by CPU manufacturers, and currently, there are two technology solutions on the market: Intel-VT and AMD-V. KVM was first integrated into the Linux kernel version

Fig. 3.10. KVM logo.

2.6.20, introduced in RHEL 5.4, and officially released on February 5, 2007. KVM can be used as long as the hardware supports Intel-VT or AMD-V. Users can determine the current hardware platform's support by using the command grep -E "vmx|svm" /proc/cpuinfo, as shown in Figure 3.11.

If the system already supports VMX (VM Extension, provided by Intel) or SVM (Secure VM, provided by AMD), users can load the corresponding Linux kernel driver to use KVM. For Intel platforms, load kvm-intel.ko; for AMD platforms, load kvm-amd.ko. As shown in Figure 3.12, the hardware platform used in this book is Intel, so the kernel displays as kvm_intel.

However, KVM alone is not enough. It only provides the interface for virtualization and does not have corresponding commands and graphical tools. It can only be controlled through specified interface APIs. Therefore, user-level commands or graphical tools are also needed. Currently, there are many tools that can be used, with the most commonly used being QEMU, VirtualBox, and VMware. KVM is a full virtualization technology solution, so users do not need to modify any Linux or Windows images to run simultaneously, and they are independent and not affected by each other.

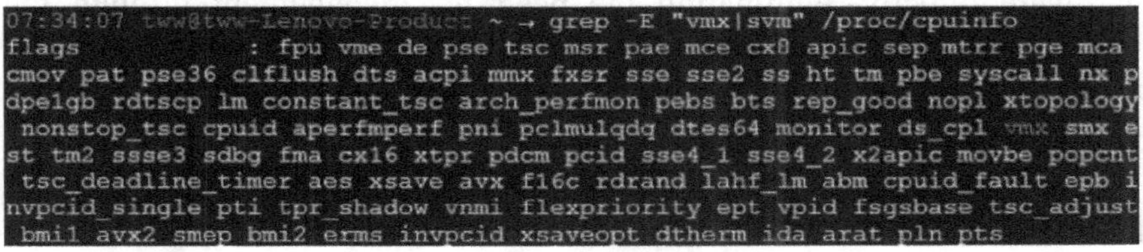

Fig. 3.11. Command to determine if the CPU supports hardware virtualization.

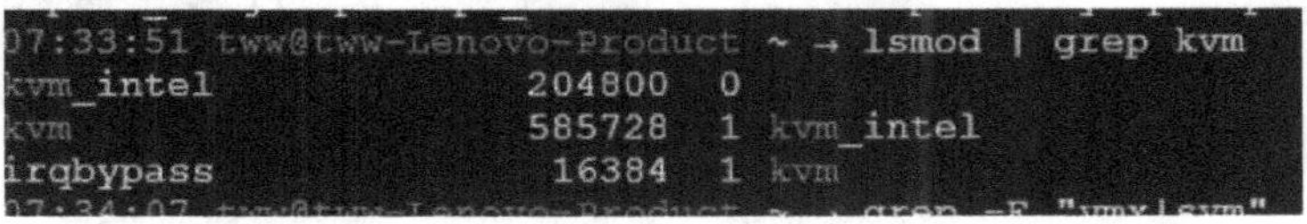

Fig. 3.12. Loading status of KVM modules.

3.3.3. *Hyper-V*

Hyper-V (https://docs.microsoft.com/en-us/virtualization/hyper-v-on-windows/about/) is a virtualization technology introduced by Microsoft, initially built into Windows Server 2008. Like VMWare ESXi and Xen, it adopts a bare-metal architecture, running directly on top of the hardware; its logo is shown in Figure 3.13.

The purpose of Hyper-V is to provide a more familiar and cost-effective virtualization infrastructure software for a wide range of users, thereby reducing operational costs, increasing hardware utilization, optimizing infrastructure, and improving server availability. Hyper-V adopts a microkernel architecture, balancing the requirements of security and performance. Due to the small code size of the Hypervisor at the core of Hyper-V, which does not include any third-party drivers, it is highly streamlined, secure, reliable, and efficient, making full use of hardware resources to bring VM system performance closer to that of real systems. Hyper-V adopts a high-speed memory bus architecture based on VMBUS (VM Bus). All hardware requests from VMs, whether they are for graphics cards, mice, keyboards, or other devices, can be directly sent through the VMBUS bus to the root partition's Virtualization Service Provider (VSP) via the Virtualization Service Consumer (VSC) for virtualization services. The VSP then calls the corresponding device driver to access the hardware directly, eliminating the need for management through the Hypervisor. In this way, hardware requests from each VM no longer need to go through multiple user mode and kernel mode context switches, greatly improving operational efficiency.

It is also possible to run a Linux operating system in Hyper-V. with the installation components related to Linux. These components can be a Linux kernel that supports Xen or specially designed integrated components for Linux. These components already include the necessary drivers, so users do not need to worry about driver-related issues in Hyper-V. After installing these components, Hyper-V can perfectly support Linux.

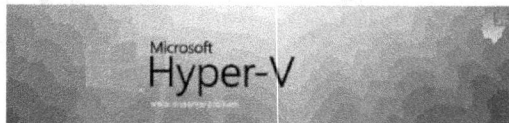

Fig. 3.13.   Hyper-V logo.

Hyper-V can create VMs using two simulation methods: paravirtualization and full virtualization. Paravirtualization requires the VM to have the same operating system as the physical host (usually the same version of Windows) to achieve high performance; full virtualization requires CPU support for full virtualization features, such as Intel VT or AMD-V, to create VMs using different operating systems, such as Linux or Mac OS.

3.3.4. *VMware*

VMware (https://www.vmware.com/) has a variety of virtualization products, catering to different needs with different product series, such as VMware ESX/ESXi, VMware Workstation, and VMware Player.

(1) *VMware ESXi*: VMware ESXi, formerly known as VMware ESX (Elastic Sky X), is an enterprise-level virtualization product, with its logo shown in Figure 3.14. VMware ESXi is a virtualization product developed to manage and run guest VMs directly on bare metal (hardware). VMware ESXi is not software that can be installed within an operating system; it includes and integrates the necessary operating system components, such as the kernel. Starting from version 4.1, ESX was officially renamed ESXi. ESXi replaces the related components of ESX, and the replaced components resemble a complete operating system. Currently, ESX/ESXi holds a significant position in the VMware virtualization infrastructure software suite.

ESXi is different from other VMware products in that it runs directly on bare metal (without an operating system) and has its own kernel. During boot-up, it first launches the Linux kernel, which is then followed by loading a series of special virtualization components, including the core module of ESX itself (VMkernel). The primary VMs in ESXi are based on the Linux kernel and are initially started by the built-in service console of ESXi. During normal

Fig. 3.14.   Logo of VMware ESXi.

operation, VMkernel takes control of hardware and runs on it. However, starting from version 4.1, ESXi no longer integrates the Linux kernel. Instead, it has updated VMkernel to be more like a complete microkernel, providing various interfaces including those for hardware, customer operating systems, and service consoles.

(2) *VMware WorkStation*: VMware Workstation is a powerful desktop virtualization software that cannot run on bare metal and requires the support of an operating system. Therefore, it is a hosted architecture type of virtualization product. VMware Workstation allows users to run different operating systems simultaneously on a single desktop, simulate complete network environments, and manage multiple VMs. Its logo is shown in Figure 3.15.

(3) *VMware Player*: VMware Player has the same functionality as VMware Workstation, but it is free. VMware Player can only run one VM at a time, but it can manage multiple VMs.

3.3.5. *Xen*

Xen (https://www.xenproject.org/) is an open-source project from the University of Cambridge and is the earliest open-source virtualization engine. It is now developed by the Linux Foundation, supported by Intel. Its logo is shown in Figure 3.16. Xen adopts a bare-metal architecture, running directly on hardware and using a microkernel implementation. It supports running multiple different operating system instances concurrently on the same device. Xen supports IA-32, x86-64, and ARM platforms. Currently, Xen is the only open-source virtualization engine with a bare-metal architecture on the market. It is most commonly used for server virtualization and Infrastructure as a Service (IaaS).

Fig. 3.15.   Logo of VMware.

Fig. 3.16. Xen logo.

The following are some of the significant characteristics and advantages of Xen:

(1) The kernel is very small, with few interfaces. Because it is a microkernel design, it uses very little memory and has few interface data, making it more secure and stable than other virtualization architectures.
(2) Xen supports a variety of operating systems, including Windows, NetBSD, and OpenSolaris. The most commonly installed operating system on Xen is Linux.
(3) Driver isolation: The Xen framework allows the main device drivers in the system to run within the VM itself. If one of the drivers malfunctions, it only requires a restart of the VM running the driver or a restart of the corresponding driver module within the VM, without affecting other systems running on the system. They are all independent and do not affect each other.
(4) Paravirtualization: Xen uses paravirtualization, where the corresponding Guest OS needs to be modified and adjusted, which can significantly improve its efficiency compared to fully virtualized operation. Additionally, it can run on hardware devices that do not support virtualization. Lastly, Xen also supports full virtualization, but only hardware-assisted full virtualization, meaning the hardware needs to support Intel-VT or AMD-V.

Xen currently runs on machines with x86 architecture and requires a P6 or newer model CPU (such as Pentium Pro, Celeron, Pentium II, Pentium III, Pentium IV, Xeon, AMD Athlon, and AMD Duron) to operate. Xen supports multiprocessors and simultaneous multithreading (SMT). Known for its high performance and low resource consumption, Xen has gained high recognition and strong support from many world-class software and hardware vendors such as IBM, AMD, HP, Red Hat, and Novell. It has been widely used by numerous enterprises and organizations at home and abroad to build high-performance virtualization platforms.

3.3.6. *Docker*

Docker (https://www.docker.com/) was originally an internal project initiated by Solomon Hykes, the founder of the dotCloud company. Docker is an innovation based on dotCloud's years of cloud service technology and was open-sourced under the Apache 2.0 license in March 2013. Its main project code is maintained on GitHub. Docker later joined the Linux Foundation and established the Open Container Initiative. The logo of Docker is shown in Figure 3.17.

Docker is an open-source application container engine, where instances running inside containers are isolated from each other, belonging to a form of operating system virtualization. Docker allows developers to package applications and dependencies into a portable container, which can then be deployed on any popular operating system. Importantly, these containers are not dependent on any specific language, framework, or system. An image (packaged file) is a lightweight, standalone executable package that includes all the dependencies it needs to run, such as software, libraries, environment, and configuration files. A container, on the other hand, is a running instance of an image, meaning the image is loaded into memory and executed. This running image is completely isolated from the host environment, except for accessing host files and ports, having no other relationship with the host. In simple terms, containers are like sandboxes, where each sandbox's operation does not affect others, and the system's other processes do not interfere with the sandboxes. They have no interfaces with each other and are independent. While sandboxes are related to the system, Docker's containers are system-agnostic, as the

Fig. 3.17. Docker logo.

packaged image contains all the necessary dependencies, allowing for universal use across all Docker environments with just one packaging. Similarly, Docker is easy to use and deploy, with minimal performance overhead and easily deployable to local or data center environments. Since containers run directly on the local host kernel, their efficiency is higher compared to running within VMs, as each container runs as a separate process and consumes less memory than VMs. The comparison between containers and VMs is shown in Figure 3.18.

For a VM framework (as shown in the right part of Figure 3.18), running applications from different operating systems is feasible on this framework. Because VMM can simulate the corresponding hardware for each VM as needed, install the appropriate operating system, and then install and run the application within this operating system. Due to the characteristics of VMs, applications of any type can be run.

For containers (as shown in the right part of Figure 3.18), since they all run on the same kernel framework, they cannot run applications from different operating systems. For example, running Linux and Windows applications simultaneously is not supported by Docker because it does not support running containers of different types simultaneously, which have Windows and Linux operating systems, respectively. If the HOST OS is Linux, Docker can start multiple containers related to the Linux kernel and run Linux-related applications in these containers. If the HOST OS is Windows, it can start multiple containers related to Windows and run Windows-related applications in these containers. Containers do not require VMs, so they use less memory and run faster.

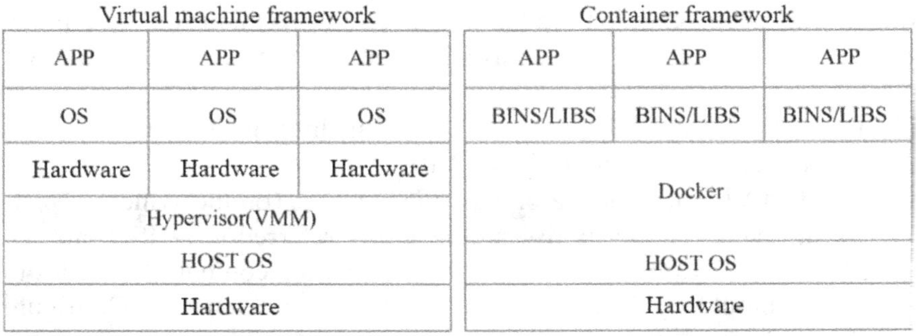

Virtual machine framework			Container framework		
APP	APP	APP	APP	APP	APP
OS	OS	OS	BINS/LIBS	BINS/LIBS	BINS/LIBS
Hardware	Hardware	Hardware	Docker		
Hypervisor(VMM)					
HOST OS			HOST OS		
Hardware			Hardware		

Fig. 3.18. Comparison between VMs and containers.

Therefore, as can be seen from Figure 3.18, one difference between the VM framework and the container framework is that VMs can run directly on bare-metal architecture, while Docker can only be built based on the HOST OS. VMM simulates physical hardware devices, while Docker does not need to, as all containers share the physical resources and kernel of the HOST OS. Therefore, applications running in Docker are strongly related to the HOST OS, while applications running in VMs managed by VMM have no relationship with the HOST OS, and the HOST OS is optional.

3.4. Application and Practice of Common Virtualization Technologies

In the previous sections, we briefly introduced some concepts, classifications, technologies, and principles of VMs. In this section, we explain the use of some common virtualization technologies, including setting up virtualized environments, cloning VMs, and taking snapshots of VMs.

3.4.1. *Virtualization environment setup*

To provide a more intuitive explanation of the purpose and principles of virtualization, this book uses Type 2 server virtualization technology with a hosted architecture as an example. The demonstration environment used in this book includes VMM and VMware Workstation. The guest VM operates on the Debian-9.1.0 operating system. Note that if you are using a different version, the installation steps may vary slightly. Please refer to the official documentation for the corresponding version for more information. The steps to set up the virtualization environment are as follows:

(1) Launch VMware Workstation. After launching it, its main interface will appear, as shown in Figure 3.19.
(2) Create a VM and select a prepared ISO image (on the Windows platform, images ending in .iso are generally referred to as ISO images. On other platforms, the user needs to check the content of the files to confirm, but they usually also end with the .iso extension). Configure the VM name as Debian and start the VM. These operations are shown in Figure 3.20.

Fig. 3.19. Main interface of VMware Workstation.

Fig. 3.20. Starting the VM.

(3) After starting, the installation interface that appears is as shown in Figure 3.21. Select the default graphical installation option "Graphical install", as shown in the figure.

After installation, if the system displays an interface as shown in Figure 3.22, it means that the client host installation is complete.

At this point, the virtualized environment setup is complete, and users can start using the VM.

3.4.2. *Clone VM*

Cloning, in simple terms, means copying. Cloning a VM can speed up the setup of the testing environment and accelerate the release of versions. Developers often deploy the new version in advance in the VM environment for the version to be released. Testers can either directly use the environment deployed by developers or only modify a small part of the configuration, such as the IP address, which greatly reduces the workload of testers in setting up the environment and saves time and costs. This

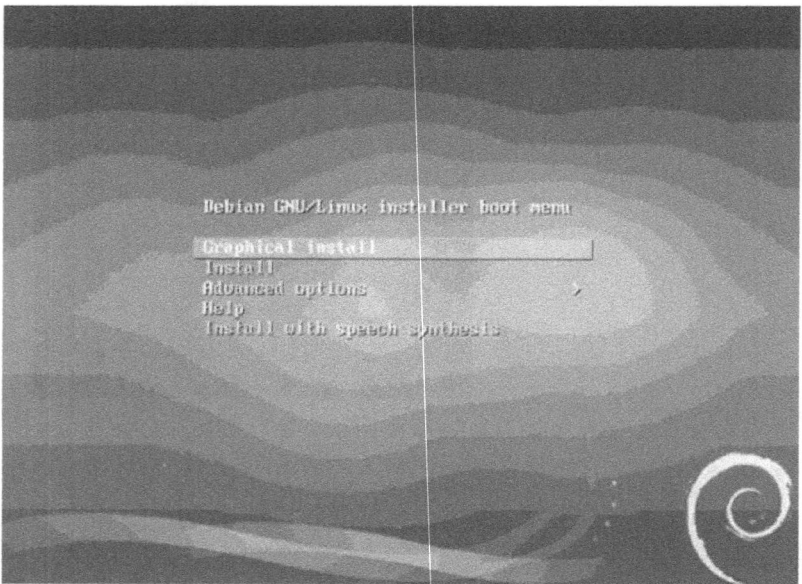

Fig. 3.21.   Graphical installation.

Fig. 3.22. Installation completion interface.

section describes how to clone a VM. Note that cloning a VM can only be done when the VM is powered off, that is, when it is shut down.

(1) Start the VM. In the VMware Workstation management window, select: VM → Manage → Clone, as shown in Figure 3.23.

(2) After selecting Clone, a dialog box will pop up, as shown in Figure 3.24, for selecting the clone type. In this section, we choose to perform a full clone. Either of the two types (linked clone or full clone) can be selected here. A linked clone is a reference to the original VM, requiring less storage space; a full clone is a complete copy of the current state of the original VM, requiring more storage space. For beginners, choosing the second clone type is more convenient, as it allows the cloned files to be easily copied to other locations.

(3) After selecting "Create a full clone" mode, click the "Next" button; a dialog box will pop up, as shown in Figure 3.25, displaying the progress of cloning the VM.

(4) After the cloning is completed, the VM will appear as shown in Figure 3.26. Then, the user can proceed to test the cloned

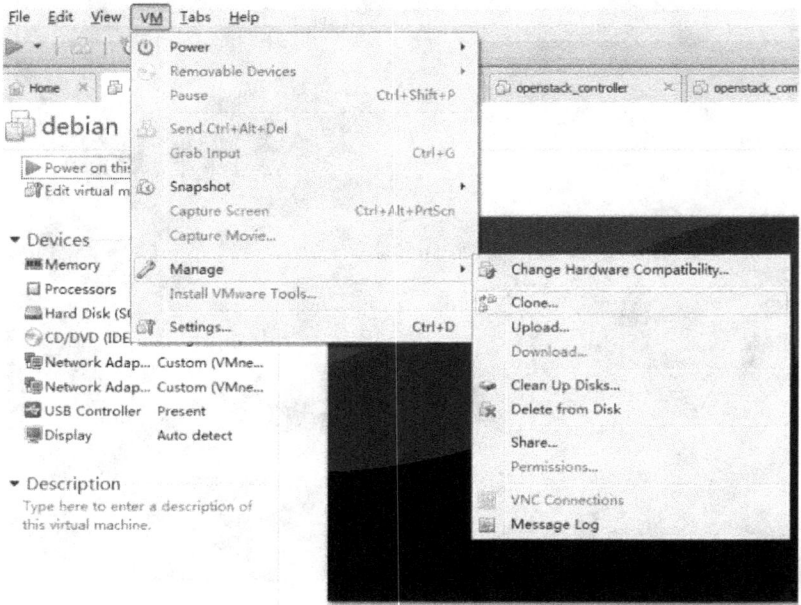

Fig. 3.23.   Select clone.

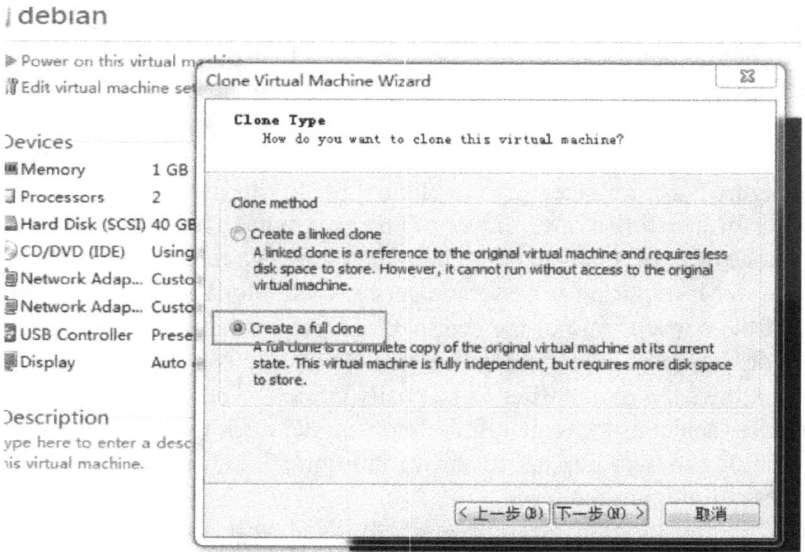

Fig. 3.24.   Select the clone type.

debian

▶ Power on this virtual machine
⚙ Edit virtual machine set

Clone Virtual Machine Wizard

Cloning Virtual Machine

▾ Devices

📠 Memory	1 GB
🖵 Processors	2
💾 Hard Disk (SCSI)	40 GB
💿 CD/DVD (IDE)	Using
📶 Network Adap...	Custo
📶 Network Adap...	Custo
🔌 USB Controller	Prese
🖥 Display	Auto

✔ Preparing clone operation

VMware Workstation

Cloning...

[Cancel]

▾ Description

Type here to enter a desc
this virtual machine.

[Close]

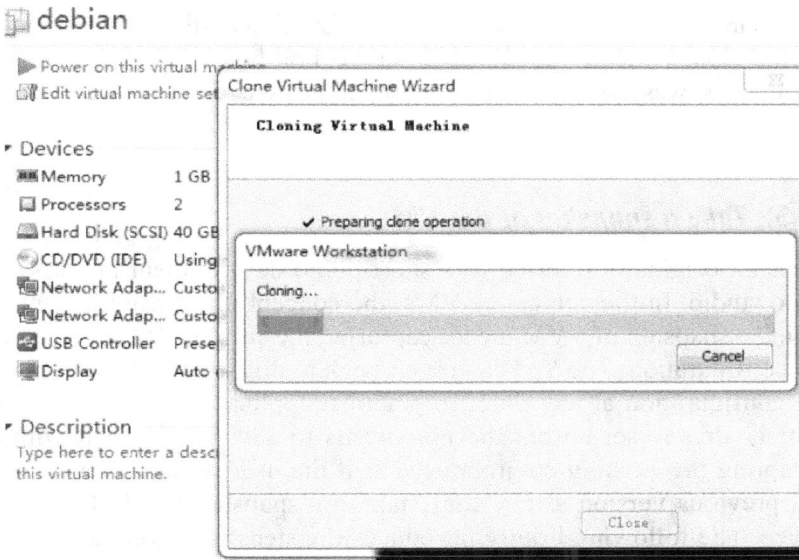

Fig. 3.25. Cloning in progress.

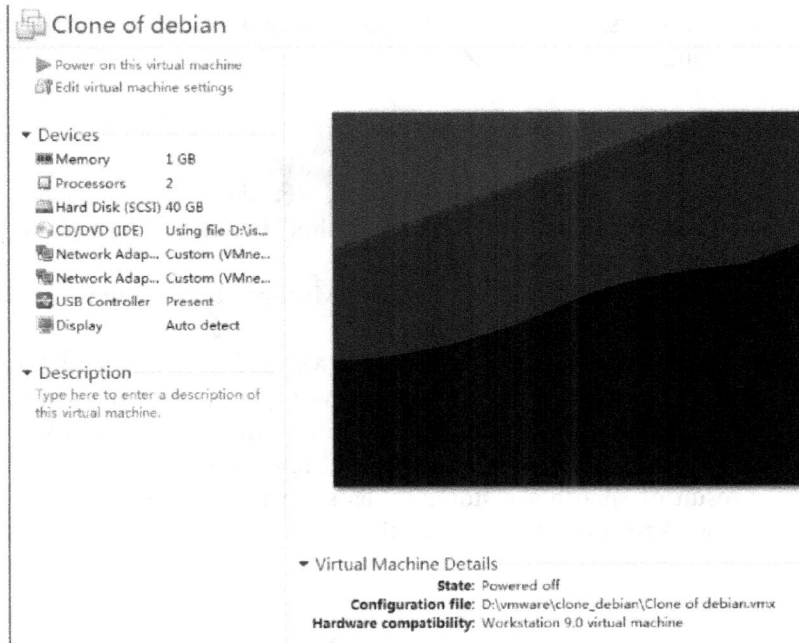

Clone of debian

▶ Power on this virtual machine
⚙ Edit virtual machine settings

▾ Devices

📠 Memory	1 GB
🖵 Processors	2
💾 Hard Disk (SCSI)	40 GB
💿 CD/DVD (IDE)	Using file D:\is...
📶 Network Adap...	Custom (VMne...
📶 Network Adap...	Custom (VMne...
🔌 USB Controller	Present
🖥 Display	Auto detect

▾ Description

Type here to enter a description of
this virtual machine.

▾ Virtual Machine Details

State: Powered off
Configuration file: D:\vmware\clone_debian\Clone of debian.vmx
Hardware compatibility: Workstation 9.0 virtual machine

Fig. 3.26. Clone completed.

environment. At this point, it will be found that all configurations of the cloned VM are identical to the version before cloning. If the IP address was fixed before cloning, it is essential to update the IP address after cloning to avoid potential communication issues.

3.4.3. *Take a snapshot of the VM*

Snapshot originally referred to a short photo development process in a photo studio. In the context of VMs, the concept of snapshot is similar. Taking a snapshot of a VM means capturing the current VM environment and configuration. The VM can revert back to this snapshot environment and configuration at any time. In practical applications, if the environment is already set up but the user wants to add new content without disrupting the existing environment, or if the user wants to revert back to a previous version at any time, taking a snapshot would be a good choice. The following briefly introduces the steps for taking a snapshot of a VM. Note that taking a snapshot is independent of the VM's running state. If a snapshot is taken while the VM is running, it will be in a running state when restored. If a snapshot is taken while the VM is powered off, it will remain powered off after restoration. To demonstrate the effect of snapshots, this section considers taking a snapshot while the VM is running.

(1) Start the VM, select VM → Snapshot → Take Snapshot, as shown in Figure 3.27.
(2) In the "Take Snapshot" dialog box, set the snapshot name to "Snapshot1", and click the "Take Snapshot" button to take a photo, as shown in Figure 3.28.
(3) Select VM → Snapshot → Snapshot Manager to view the snapshot results, as shown in Figure 3.29.
(4) Perform a series of operations on the VM, as shown in Figure 3.30.
(5) Select VM → Snapshot → Snapshot Manager, choose the snapshot named "snapshot1", and then click on the "Go To" command to perform snapshot restoration, as shown in Figure 3.31.
(6) The result of snapshot restoration, as shown in Figure 3.32, reveals that the VM environment at this moment is identical to the

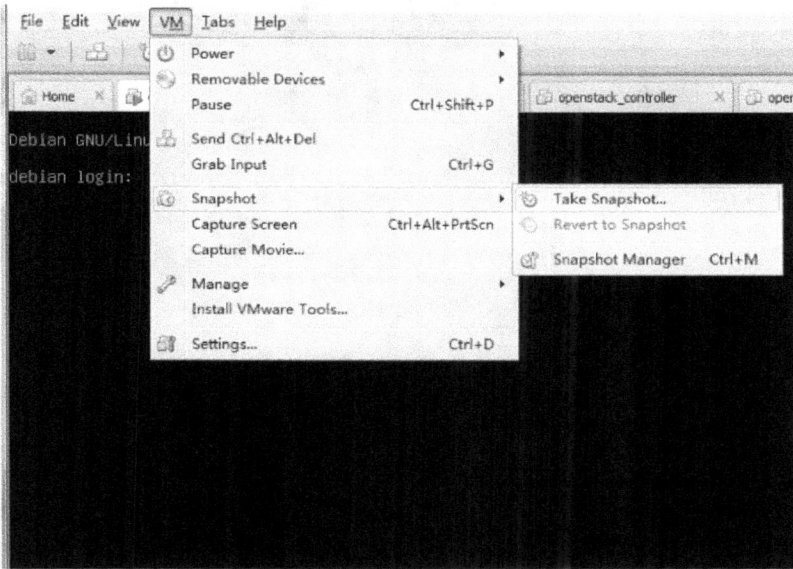

Fig. 3.27. Select take snapshot.

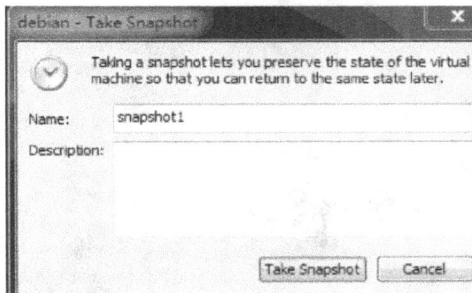

Fig. 3.28. Setting snapshot name.

environment when the snapshot was initially taken with no changes. Note that users can take snapshots of the VM at any time and restore the results of any snapshot. Readers can practice and experiment with this on their own.

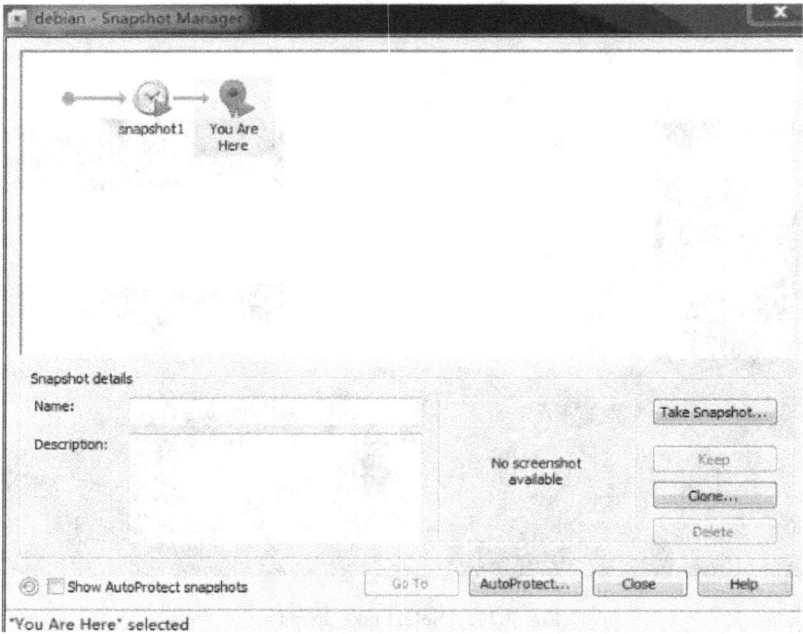

Fig. 3.29.　Snapshot management.

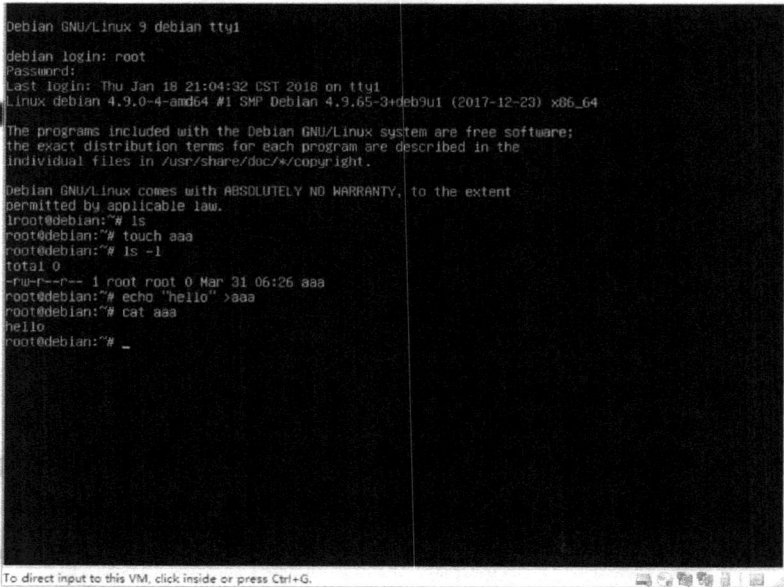

Fig. 3.30.　Operating the VM.

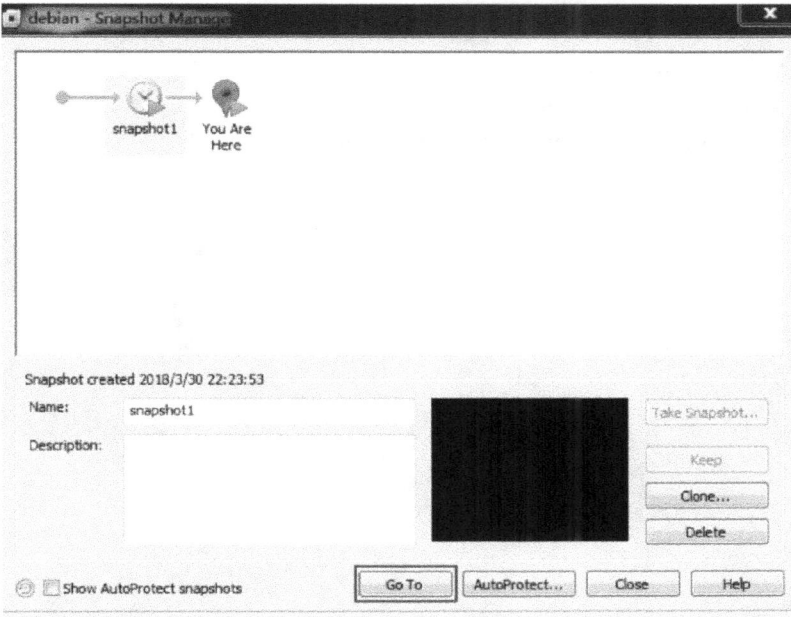

Fig. 3.31. Snapshot restoration.

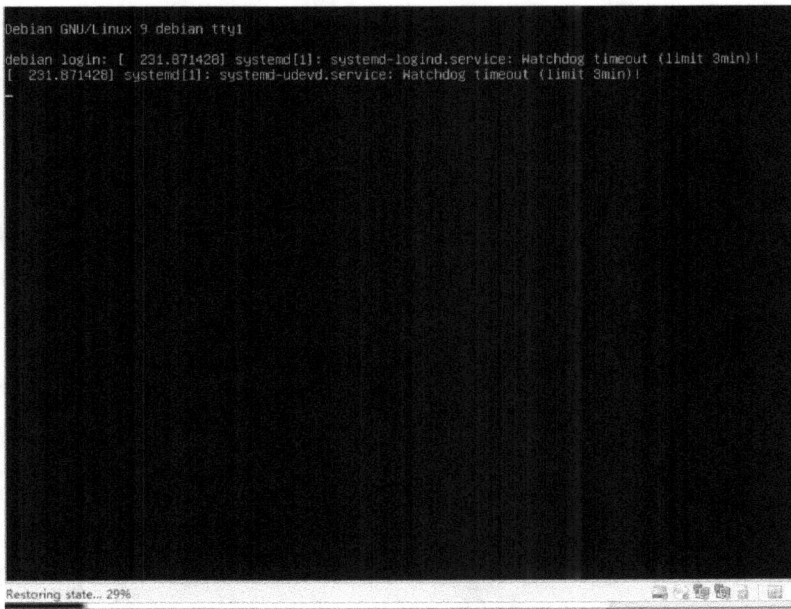

Fig. 3.32. Snapshot restoration status.

Exercises

(1) What is virtualization technology?
(2) Why use virtualization technology?
(3) What are the common virtualization technologies?

Chapter 4

Data Centers and Cloud Storage Technology

Nowadays, the explosion of various types of data has forced enterprises to continuously enhance their data processing capabilities based on data centers. At the same time, cloud computing technology is constantly bringing new momentum to the development of data centers and is changing the traditional models of data centers.

Cloud-based data storage technology, known as cloud storage, is an extended concept of computer storage. Cloud storage integrates distributed system files and, based on network technology, centrally manages computer storage devices and storage software through cluster applications to achieve collaborative cooperation. Cloud storage is powerful, and its data storage mode is innovative. The key to its application is software rather than hardware. It ensures the service functionality transformation of computer storage devices through the organic combination of different devices. Therefore, cloud storage has high transparency in wide area networks (WANs), provides convenient data access, and has a wide range of business access. Cloud storage is accessed through web service Application Programming Interfaces (APIs) or web-based user interfaces.

This chapter introduces the concepts of data centers and cloud storage, the relationship between cloud storage and cloud computing, and the applications and development trends of cloud storage.

4.1. The Concept of Data Centers

A data center is a specific network of devices for global collaboration, used to transmit, accelerate, display, compute, and store data information on the Internet as a network infrastructure. A data center is not merely a network concept but also a service concept. It forms part of the fundamental network resources. The establishment of data centers enables businesses and individuals to quickly carry out operations over the network, focusing on their core activities and reducing concerns about their IT aspects. Data centers are the infrastructure for cloud computing, which offers various services based on data centers.

4.1.1. *Definition, role, and classification of data centers*

A data center is a complex set of facilities, which includes not only computer systems and other associated network, storage, and related devices but also redundant data communication connections, environmental control devices, monitoring equipment, and various security devices. Google, in its publication *The Data Center as a Computer*, defines a data center as follows: a multifunctional building that can house multiple servers and communication devices. These devices are placed together because they share the same environmental requirements and physical security needs, and such an arrangement facilitates maintenance, rather than being merely a collection of servers.

Figure 4.1 shows a typical data center topology, illustrating that a data center comprises several fundamental components, including servers, networks, and storage. Storage, depending on different scenarios, includes Network Attached Storage (NAS, providing a file-sharing space for unstructured data) and Storage Area Network (SAN, allowing block data access, typically used for storing structured data, such as databases, emails, and other business application data). Enterprise-level storage usually also encompasses backup or disaster recovery systems, and the data center shown in Figure 4.1 includes dedicated backup storage.

After the emergence of cloud computing, new challenges and requirements have been proposed for data centers. However, they still comprise the three basic components: servers, network, and storage. The form and control methods of these components have undergone significant changes, such as the Software Defined Data Center (SDDC) and Software Defined Storage (SDS). SDDC can be considered as the optimal implementation

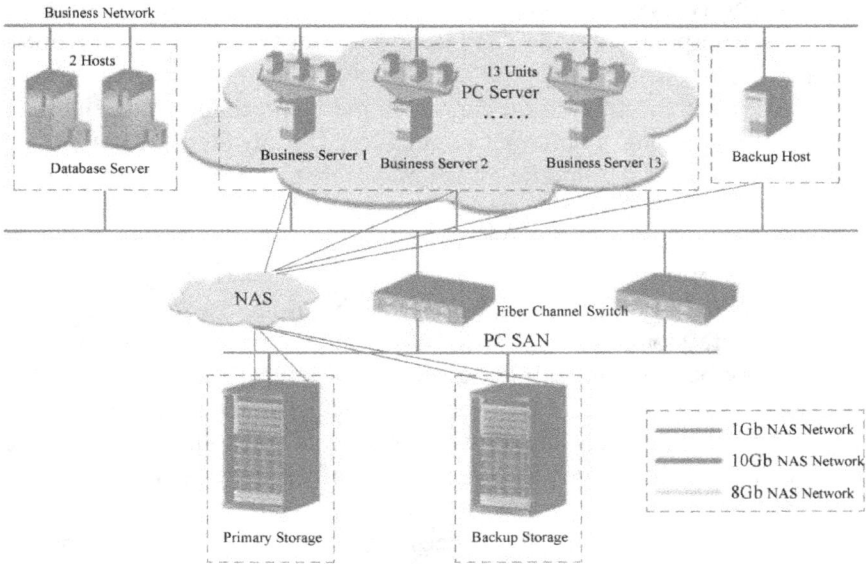

Fig. 4.1. Typical data center topology diagram.

of a private cloud computing platform, with SDS being an important component within it.

Currently, IT giants, including Google and Microsoft, have invested heavily in the construction of data centers. Based on the role of data centers, they can be classified into two categories: Enterprise Data Centers (EDCs) and Internet Data Centers (IDCs).

- EDCs achieve the ultimate goal of data sharing and utilization by implementing unified data definitions and naming conventions, as well as centralized data environments. For a specific business or entity, the data center is a combination of business data storage technology and a data warehouse. However, some entities, like research institutions, may only have a data warehouse since they do not process business transactions but have analysis needs. EDCs are categorized by size into departmental data centers, corporate data centers, IDCs, and co-located data centers. Through these data centers, which range from small to large, businesses can run various applications. A typical corporate data center's equipment usually include mainframe devices, data backup devices, data storage devices, high availability systems, data security systems, and database systems. These components need to be placed together to ensure they can operate as a whole.

- IDCs differ from the usual corporate and institutional data centers in that they belong to Internet companies. Due to the vast amount of information on the Internet, it is impossible to store all information in a data center's database, and a data center cannot process that much information either. Thus, the role of a data center in an Internet environment is to enhance the processing speed and effectiveness of Internet data. IDCs provide large-scale, high-quality, secure, and reliable professional services such as server hosting, space leasing, and network bandwidth for Internet content providers, enterprises, media, and various websites. IDCs serve as hosting facilities for merchants, enterprises, or a cluster of website servers. It is the infrastructure architecture that electronic commerce of all types depends on for secure operations. It also serves as a secure platform supporting enterprises and their business alliances (including distributors, suppliers, and customers) in implementing value chain management.

4.1.2. *Development trends of data centers in the era of cloud computing and big data*

(1) *Scalability: Large Data Centers Gain Market Preference*: In recent years, the construction scale of data centers has been expanding continuously, with many mega data centers planned to occupy hundreds of thousands of square meters. From a market acceptance perspective, the data center industry is undergoing a reshuffling, with users preferring to choose service providers that have strong technological capabilities and superior service systems. For instance, the future data centers of Google, Microsoft, Tencent, and Alibaba are moving toward globalization, internationalization, and scalability.

(2) *Virtualization: Traditional Data Centers to Migrate Resources to the Cloud*: In traditional data centers, servers, network devices, storage devices, and database resources are independent of one another, with no association between them. Virtualization technology has changed this state of resources being unrelated across different data centers. With the deepening application of virtualization technology, server virtualization has moved from concept to practice, gradually

expanding into the application domain. In the future, more applications will migrate to the cloud based on virtualization technology. Therefore, traditional data centers will evolve into cloud data centers.

(3) *Greening: Traditional Data Centers to Transition to Green Data Centers*: The continuously rising energy costs and growing computing demands are making the energy consumption of data centers an increasingly prominent issue. The construction process of data centers adheres to the basic construction guidelines of saving land, water, electricity, and materials and protecting the environment. Energy saving, environmental protection, green, and low carbon will undoubtedly become the themes of the next generation of data center construction.

(4) *Centralization: Traditional Data Centers to Enter a Phase of Integration and Reduction*: The current state of dispersed office environments has led to a scattered layout of application systems. However, there is a realistic need for centralized processing of data from branch offices, making data center centralization a necessity. In the future, with technological advancements, the trend toward data center integration and centralization will become increasingly apparent.

(5) *Low Cost: The Value of Data Increasingly Highlighted and at Lower Costs*: Virtualization technology has improved resource utilization rates, simplified the management dimensions of data centers, and effectively saved maintenance costs. While reducing costs, it also realizes the maximization of data value.

4.2. Cloud Storage Overview

In the era of big data, the massive amount of data generated has a large base volume and complex data formats, requiring higher real-time and efficient data processing capabilities. Traditional data storage technologies are fundamentally incapable of meeting these demands. Therefore, there is a need for rapid improvement in data storage technologies to provide more powerful, stable, and secure storage performance for cloud computing, leading to the development of cloud storage technology.

4.2.1. *Concept of cloud storage*

The primary use of data storage technology in its application is temporary files, which are a form of data stream formed during processing.

By searching basic information, data records are formatted and stored in both external and internal computer storage media. Data storage requires naming according to related information features, reflecting flowing data as a data stream in the system, simultaneously presenting static and dynamic data characteristics.

Cloud storage is an online storage model where data are stored in multiple virtual servers hosted by a third party rather than on dedicated servers. Hosting companies operate large data centers and offer or lease storage space to those needing data storage services; based on the customers' needs, data center operators prepare storage virtualization resources at the backend and provide them to customers as a storage resource pool (Storage Pool), allowing them to store files or objects. In fact, these resources may be distributed across numerous server hosts.

Technically, cloud storage refers to the technology that, through cluster, network, or distributed technologies, combines a large amount of storage devices of various types in the network via application software to work together, providing collective data storage and business access functions. Based on cloud storage technology, data are stored on a variety of virtual servers, usually managed by third-party organizations rather than dedicated servers. When the core of a cloud computing system's operation and processing is the storage and management of a large amount of data, the system needs to be equipped with numerous storage devices, transforming the cloud computing system into a cloud storage system centered on data storage and management.

The design philosophy of cloud storage technology is to minimize and decentralize, breaking down big data across various storage servers for storage, then linking these servers through the network to work in collaboration, forming a large cluster system, which is accessed via virtualization technology on the cloud storage system. Therefore, cloud storage is a network-based data storage model, an extension and development from the concept of cloud computing.

From the above description, it can be seen that cloud storage differs from traditional storage systems in the following ways: From a functional requirement perspective, compared to the singular function of traditional storage, cloud storage systems are more open and diversified; in terms of data management, cloud storage handles more data types and larger volumes of data.

4.2.2. *Structure of cloud storage systems*

Cloud storage systems are unlike traditional storage devices, which are simply hardware. A cloud storage system is a complex system that includes hardware components, such as switches, routers, network adapters, fiber optic cables, relays, amplifiers, and repeaters, and a vast array of software systems, including operating systems, system software, tool software, and application software. Centered on storage devices, cloud storage systems can connect to external storage and handle access demands through necessary application software interfaces. The structure of a cloud storage system is composed of four layers: the storage layer, the basic management layer, the application interface layer, and the access layer, as shown in Figure 4.2.

(1) *Storage Layer*: The storage layer is the most fundamental part of a cloud storage system. Storage devices can be fiber channel (FC) storage devices, IP storage devices such as NAS and ISCSI, or DAS storage devices such as SCSI or SAS. The storage devices in cloud

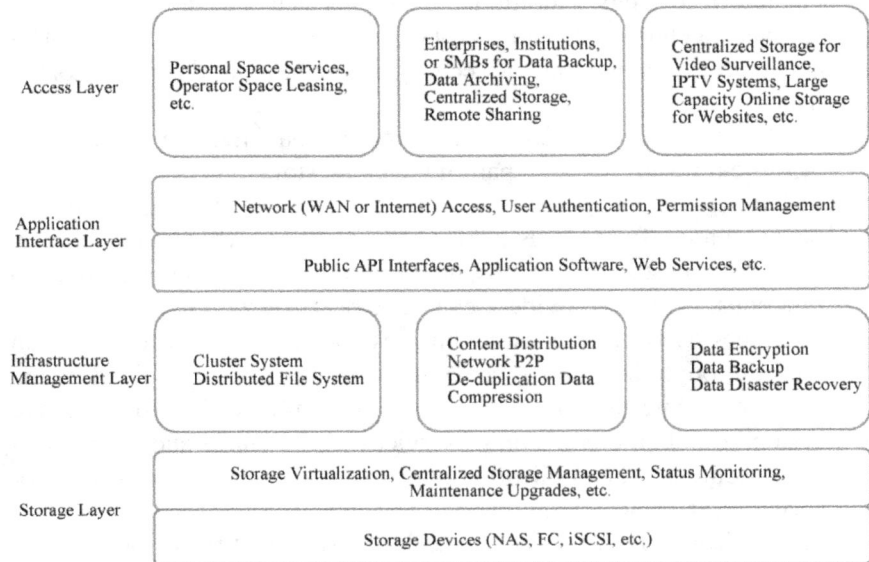

Access Layer	Personal Space Services, Operator Space Leasing, etc.	Enterprises, Institutions, or SMBs for Data Backup, Data Archiving, Centralized Storage, Remote Sharing	Centralized Storage for Video Surveillance, IPTV Systems, Large Capacity Online Storage for Websites, etc.

Network (WAN or Internet) Access, User Authentication, Permission Management

Application Interface Layer

Public API Interfaces, Application Software, Web Services, etc.

Infrastructure Management Layer	Cluster System Distributed File System	Content Distribution Network P2P De-duplication Data Compression	Data Encryption Data Backup Data Disaster Recovery

Storage Virtualization, Centralized Storage Management, Status Monitoring, Maintenance Upgrades, etc.

Storage Layer

Storage Devices (NAS, FC, iSCSI, etc.)

Fig. 4.2. Architecture of a cloud storage system.

storage are often numerous and distributed across different regions, connected through WANs, the Internet, or FC networks.

Cloud storage systems offer a variety of storage services, with all the data from these services stored uniformly in the cloud storage system, forming a massive data pool.

Above the storage devices is a unified storage device management system that can realize the logical virtualization management of storage devices, multi-link redundancy management, and hardware status monitoring and fault maintenance.

(2) *Basic Management Layer*: The basic management layer is the core part of cloud storage and the most challenging to implement within the cloud storage system. It enables the coordination among multiple storage devices in the cloud storage system, allowing them to provide the same type of service and offer larger, stronger, and better data access performance.

Content Delivery Network (CDN) enables users to obtain the required content nearby, alleviating Internet congestion and enhancing the response speed of website access. Data encryption technology ensures that data in cloud storage systems cannot be accessed by unauthorized users. Additionally, various data backup and disaster recovery technologies and measures prevent data loss in cloud storage, thereby maintaining the security and stability of the cloud storage itself.

(3) *Application Interface Layer*: The application interface layer is the most flexible and variable part of the cloud storage system. The application service interfaces developed for users by the cloud storage platform are referred to as public API interfaces, which include interfaces for data storage services, public resource usage, and data backup functions. Service providers can develop corresponding application interfaces according to users' business needs. Authorized users can log in through the web service application interfaces provided by the application interface layer from anywhere, utilize the cloud storage system to obtain cloud storage services, and manage and access system resources. The application interface layer also includes network access, user authentication, and access management functions.

(4) *Access Layer*: Through the access layer, any authorized user can log into the cloud storage system from anywhere using Internet terminal devices, following the access interfaces or methods provided by the operator to receive cloud storage services. Different cloud storage

operators can develop different application services and adapt existing ones based on actual business types, such as network disks, video on demand, video surveillance, remote data backup, and other applications.

From the structure of the cloud storage system, it is evident that cloud storage for users does not refer to a specific device but rather a conglomerate of numerous storage devices and servers. When users utilize cloud storage, they are not using a single storage device but the data access service provided by the entire cloud storage system. Thus, in strict terms, cloud storage is not storage but a service. The core of cloud storage is the combination of application software and storage devices, realized through application software that transitions storage devices into storage services.

4.2.3. *Foundations of cloud storage implementation*

From the structure of cloud storage, it is evident that a cloud storage system is a cooperative ensemble of multiple devices, applications, and services. Its realization depends on the development of various technologies.

(1) *Broadband Network*: A true cloud storage system will be a vast public system distributed across multiple regions, nationwide, and even globally. Users need to connect to cloud storage through broadband access devices such as ADSL and DDN. Only with sufficient development of broadband networks can users obtain ample data transmission bandwidth, enabling the transfer of large volumes of data, and truly benefit from cloud storage services.

(2) *Web 2.0 Technology*: The core of Web 2.0 technology is sharing. Only through Web 2.0 technology can cloud storage users achieve centralized storage and sharing of data, documents, pictures, and audiovisual content via various devices, such as PCs and mobile phones.

(3) *Application Storage*: Cloud storage is not merely about storage; it is also about applications. Application storage not only has data storage functions but also has application software functions. It can be considered a combination of servers and storage devices. The development of application storage technology can significantly reduce the number of servers in cloud storage, thereby reducing system construction

costs, minimizing single points of failure and performance bottle-necks caused by servers within the system, reducing data transmission links, and enhancing system performance and efficiency. This ensures the high efficiency and stable operation of the entire system.

(4) *Cluster Technology and Distributed File Systems (DFSs)*: From the concept of cloud storage, it is evident that a single-point storage system cannot be considered cloud storage. Cloud storage consists of multiple storage devices that require cluster and distributed technologies to enable the collaborative work of these devices. Multiple storage devices can collectively offer the same service. Without the support of these technologies, true cloud storage cannot be achieved, and the so-called cloud storage would only be a series of independent systems unable to form a cloud structure.

Cluster technology can provide relatively high gains in performance, reliability, and flexibility at a relatively low cost. A cluster is a group of independent computers interconnected through a high-speed network, forming a group and managed as a single system. When a client interacts with a cluster, the cluster acts as a single, independent server. Clusters can enhance the availability and scalability of cloud storage systems. Task scheduling is a core technology within cluster systems.

A DFS indicates that the physical storage resources managed by the file system are not necessarily directly connected to the local node; they can also be connected to non-local nodes through a computer network. The design of a DFS is based on the client/server model. A typical DFS may include multiple servers accessible by multiple users. Additionally, peer-to-peer characteristics allow some systems to act in dual roles as both client and server.

(5) *CDN, Peer-2-Peer Technology, Data Compression Technology, Deduplication Technology, and Data Encryption Technology*: The basic idea of the CDN content distribution system is to avoid possible bottlenecks and links on the Internet that may affect data transmission speed and stability, making content transmission faster and more stable. By placing node servers throughout the network, which form an intelligent virtual network layer on top of the existing Internet infrastructure, a CDN system can redirect user requests to the service node closest to the user based on real-time comprehensive information such as network traffic, connection conditions, load status of each node, and distance to the user and response time.

Peer-to-Peer (P2P) technology, also known as peer networking, utilizes the computing power and bandwidth of all participants in the P2P network to perform tasks, instead of concentrating them on a few servers. Various file-sharing software using P2P technology has already been widely used. P2P technology is also employed in data communication for real-time media services, such as Voice over Internet Protocol (VoIP).

Data compression technology refers to methods that reduce data size without losing useful information to minimize storage space, enhance transmission, storage, and processing efficiency, or reorganize data according to certain algorithms to reduce redundancy and storage space. Data compression includes lossy compression and lossless compression. In computer science and information theory, data compression, or source coding, is the process of representing information with fewer bits than the uncompressed representation according to a specific encoding mechanism. For example, encoding "compression" as "comp" allows a document to be represented with fewer bits. A popular example of compression is the ZIP file format, which is used by many computers. ZIP not only provides compression but also serves as an archiver, allowing many files to be stored in a single file.

Deduplication technology is also a form of data compression, commonly used in disk-based backup systems, aimed at reducing the storage capacity used in storage systems. It works by finding duplicate variable-sized data blocks in different positions of different files within a certain time period. Duplicate data blocks are replaced with indicators. Highly redundant datasets, such as backup data, benefit greatly from deduplication technology; users can achieve reduction ratios of 10:1 to 50:1. Moreover, deduplication technology allows users to efficiently and economically replicate backup data between different sites.

Data encryption technology is an age-old technique that involves transforming plaintext into ciphertext through encryption algorithms and encryption keys, while decryption is the process of converting ciphertext back into plaintext through decryption algorithms and decryption keys. The core of data encryption technology is cryptography. Currently, data encryption technology remains one of the most reliable methods for protecting information in computer systems.

(6) *Storage Virtualization Technology and Storage Network Management Technology*: In cloud storage, the number of storage devices is vast

and mostly distributed across different regions. Managing logical volumes, storage virtualization, and multi-path redundancy across various manufacturers, models, and even different types of storage (such as FC storage and IP storage) poses a significant challenge. To solve this problem and simplify user operations, storage virtualization technology is needed.

There are three main ways to achieve storage virtualization:

(1) *Implementing Storage Virtualization on the Server Side*: Storage virtualization implemented on the server side involves mapping mirrors to peripheral storage devices through the server. Apart from data allocation, the server does not exert any control over the peripheral storage devices. Generally, storage virtualization on the server side is achieved through logical volume management. This provides a virtual layer that maps physical storage to logical volumes. The server only needs to handle logical volumes without managing the physical parameters of the storage devices.

(2) *Implementing Storage Virtualization on the Storage Device Side*: Another method of implementing storage virtualization is to virtualize the storage devices themselves. Virtualized storage devices on the storage subsystem side are typically connected to servers through large-scale RAID subsystems and multiple I/O channels. Intelligent controllers provide LUN access control, caching, and other management functions, such as data replication. The advantage of this approach is that storage device managers have complete control over the devices. Additionally, by separating storage management from the server systems, you can isolate storage management from various server operating systems, and it becomes easier to adjust hardware parameters.

(3) *Implementing Storage Virtualization on the Network Device Side*: Storage virtualization implemented on the network device side involves mapping logical volumes to peripheral storage devices through the network. Apart from data allocation, this method does not exert any control over the peripheral storage devices. Implementing storage virtualization on the network side is reasonable because it is neither on the server side nor on the storage device side, but rather in between the two environments, offering a potentially more "open" virtualization environment that can support almost any server, operating system, application, and storage device.

Generally, server-based and storage-device-based storage virtualization are preferred because these two architectures are convenient, easy to manage and maintain, relatively mature in terms of products, and have a high performance-to-price ratio.

Another issue arising from the vast number of storage devices spread over wide areas in cloud storage is the operational management of these devices. While these issues are not a concern for the users of cloud storage, they are crucial for the operating entities of cloud storage. Effective and practical means must be employed to address challenges such as centralized management difficulties, status monitoring difficulties, fault maintenance difficulties, and high labor costs. Therefore, cloud storage must possess an efficient, centralized management platform akin to network management software to achieve centralized management and status monitoring of all storage devices, servers, and network equipment in the cloud storage system. This can be achieved through storage network management technology.

4.2.4. *Characteristics of cloud storage*

As cloud data becomes increasingly rich and user data grows larger, cloud storage has become a significant focal point in the information storage field. The characteristics of cloud storage are explained as follows:

(1) *Reliability*: Cloud storage employs a storage model that divides multiple small files into several replicas to achieve data redundancy. Data are stored on multiple, different nodes, and if any node experiences a data failure, the cloud storage system automatically backs up the data to a new storage node to ensure data integrity and reliability. For large files, the system uses the Super Secure Storage coding algorithm (a method that integrates the RS algorithm into a distributed storage system to address disk space wastage caused by the simple duplication backup method used by general cloud storage support methods) to ensure data reliability. If multiple storage data nodes are damaged, the system can automatically decode and recover data using the Super Secure Storage algorithm. This algorithm is suitable for scenarios with extremely high data security requirements and effectively improves space utilization on disks. For metadata (information that describes the organization of data, data domains, and their

relationships) management nodes, a high-availability dual-machine mirroring hot backup method is used for fault tolerance. If one server encounters a problem, it can seamlessly transition to another server, ensuring that the server continuously provides service.

(2) *Security*: Cloud storage service providers often have significant financial resources, and the daily management and maintenance by numerous professional technicians can ensure the secure operation of the cloud storage system. Through strict permission management and the use of technologies such as data encryption, encrypted transmission, anti-tampering and anti-attack mechanisms, and real-time monitoring, the risk of virus and network hacker intrusion is reduced, ensuring that data are not lost and providing users with a secure and reliable data storage environment.

(3) *Ease of Management*: Once most of the data have been migrated to cloud storage, all data upgrade and maintenance tasks are handled by the cloud storage service provider, thus greatly reducing the operational and maintenance costs of enterprise storage systems. Cloud storage services offer strong scalability; when a company accelerates its development and finds that its existing storage space is insufficient, it can consider increasing the storage server capacity to meet the current business storage needs. The characteristics of cloud storage allow the service space to be easily expanded based on existing infrastructure to meet demand.

(4) *Scalability*: Scalability in storage demand (both upward and downward) can improve user costs, but this requires that cloud storage providers offer scalability for the storage itself (functional scalability) as well as for the storage bandwidth (load scalability). Another key feature of cloud storage is the geographic distribution of data (geographic scalability), which supports storing data as close as possible to users through a set of cloud storage data centers (via migration). For read-only data, they can also be replicated and distributed (using a CDN). Internally, a cloud storage architecture must be capable of scaling, and both servers and storage must be able to resize without affecting the users.

4.3. Cloud Storage and Cloud Computing

Cloud storage is an extension of the concept of cloud computing. It is a system that utilizes cluster technology, network technology, or DFSs to integrate various storage devices on the network using application software, enabling them to work collaboratively and provide data storage and business access functions to external users. In a sense, cloud storage is a cloud computing system focused on data storage and management, offering online storage services to users over the Internet. When a cloud computing system is equipped with large-capacity data storage devices, it can store and manage a vast amount of data that require computation and processing. Additionally, numerous functions are added at the basic management layer to enhance the management of stored data and ensure data security. Such a cloud computing system thus transforms into a cloud storage system.

Cloud storage saves users the expenses of equipment deployment and storage. Users only need to pay a certain fee to cloud service providers to directly enjoy the data access services of the entire cloud storage system. It can be said that cloud storage is the cornerstone of cloud computing systems.

Cloud computing technology can provide users with efficient network services capable of processing gigabytes of data within seconds. In a cloud computing system, data computation and processing are the core tasks. A large number of servers work together to process users' computational requests and deliver the results. For users, cloud storage is a service rather than merely a storage device. Compared to cloud computing, cloud storage can be viewed as a cloud computing system with large-capacity space.

4.4. Focus Areas in the Development of Cloud Storage

Cloud storage has already become a trend for the future development of storage. However, with the advancement of cloud storage technology and the integration of various search and application technologies with cloud storage, there is still a need for improvements in terms of security, portability, and data access.

(1) *Security*: For cloud storage, security remains the primary concern. Especially for cloud storage users, security is typically their foremost commercial and technical consideration. Many users have security requirements for cloud storage that exceed the security levels their

own data centers can provide. Cloud storage service providers are also striving to meet users' security demands, working hard to build data centers that are much more secure than most EDCs.

(2) *Portability*: Some users consider the portability of data when using managed storage. This is generally guaranteed, as solutions provided by some large service providers ensure that data portability can match the best traditional localized storage. Some cloud storage solutions combine strong portability features, allowing the entire dataset to be transferred to any medium chosen by the user, even to specialized storage devices.

(3) *Performance and Availability*: Past managed storage and remote storage solutions often suffered from excessive latency issues. Similarly, the very nature of the Internet itself can severely jeopardize service availability. The latest generation of cloud storage has achieved groundbreaking successes, particularly in client-side or local device caching, which keeps frequently accessed data locally, thereby effectively mitigating Internet latency issues. With local caching, even in the event of severe network disruptions, these devices can alleviate latency problems. They also enable frequently used data to respond as quickly as if it were stored locally. Using a local NAS gateway, cloud storage can even emulate the availability, performance, and visibility of terminal NAS devices while safeguarding data remotely. As cloud storage technology continues to evolve, vendors will keep striving for capacity optimization and WAN optimization to minimize data transmission latency.

(4) *Data Accessibility*: Current concerns about cloud storage technology revolve around whether cloud storage can provide sufficient accessibility for large-scale data requests or data recovery operations. These concerns are largely unfounded. Current vendors can transfer vast amounts of data to any type of medium and deliver it directly to enterprises at speeds comparable to performing copy-and-paste operations on local computers. Additionally, cloud storage vendors can offer a suite of components that emulate cloud storage on fully localized systems, allowing local NAS gateway devices to continue running without requiring reconfiguration. In the future, as large vendors build more regional facilities, data transfers will become even faster. Thus, even if catastrophic data loss occurs locally, cloud storage vendors can quickly re-transfer the data back to the customer's data center.

Exercises

(1) What are the stages of data center development?
(2) What are the main components of a data center?
(3) Describe the structural model of a cloud storage system.
(4) Briefly explain the prerequisites for implementing cloud storage.
(5) What are the classifications of cloud storage service systems? List some applications and briefly describe them.
(6) Briefly explain the characteristics of cloud storage.

Chapter 5

Parallel Computing and Cluster Technology

Parallel Computing, also known as High-Performance Computing (HPC), High-End Parallel Computing, or Super Computing, has rapidly developed to provide critical support for the growth of other technologies. Cloud computing focuses on two key elements: computing power and storage capacity, with computing power relying on parallel computing. Therefore, understanding the concept and classification of parallel computing is essential for learning about cloud computing. Cluster technology, the foundation of parallel computing, is also a key aspect of cloud computing. This chapter introduces the fundamental concepts and classifications of parallel computing, the cloud computing infrastructure—clusters, and parallel programming—MPI programming.

5.1. Overview of Parallel Computing

Parallel computing refers to the process of using multiple computing resources simultaneously to solve computational problems. It is an effective method for improving the computational speed and processing capacity of computer systems. The basic idea is to use multiple processors to collaboratively solve a single problem by breaking down the problem into several parts, with each part being processed by an independent processor. In computer terminology, parallelism is the ability to break down a complex problem into several sub-problems that can be processed simultaneously.

5.1.1. *Concept of parallel computing*

The primary goal of parallel computing is to speed up problem-solving by expanding the scope of the problem and tackling large, complex computational challenges. For example, large-scale serial tasks requiring extensive computation over a long period can be broken into several relatively independent modules and executed in parallel to save time as shown in Figure 5.1.

The idea of parallel computing can be understood through an analogy: If there are 21 acres of land to weed, and one person can weed only one acre per day, it would take 21 days to complete the task. However, if 21 people work together, it would only take one day to finish, greatly reducing the time required. From this example, it can be seen that task decomposition does not reduce the workload, but rather increases the labor force to save time, as shown in Figure 5.2. This is essentially the basic idea of parallel computing in addressing problems.

Parallel computing is in contrast to serial computing. Typically, serial computing refers to the execution of software operations on a single computer (with a single central processing unit), where the CPU sequentially executes a series of instructions to solve a problem, but only one instruction is available for use at any given time. Parallel computing adds processors on the basis of serial computing, allowing the simultaneous execution of multiple instructions. The difference between the two is illustrated in Figure 5.3.

Parallel computing refers to the process of using multiple computing resources simultaneously to solve computational problems. It is an effective means of enhancing the computing speed and processing capacity of

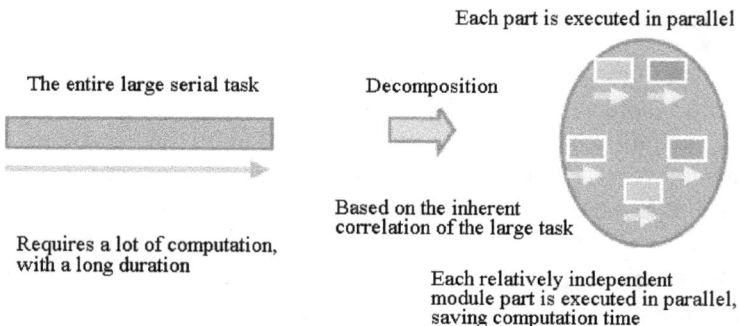

Each part is executed in parallel

The entire large serial task

Decomposition

Requires a lot of computation, with a long duration

Based on the inherent correlation of the large task

Each relatively independent module part is executed in parallel, saving computation time

Fig. 5.1.   Parallel computing task decomposition.

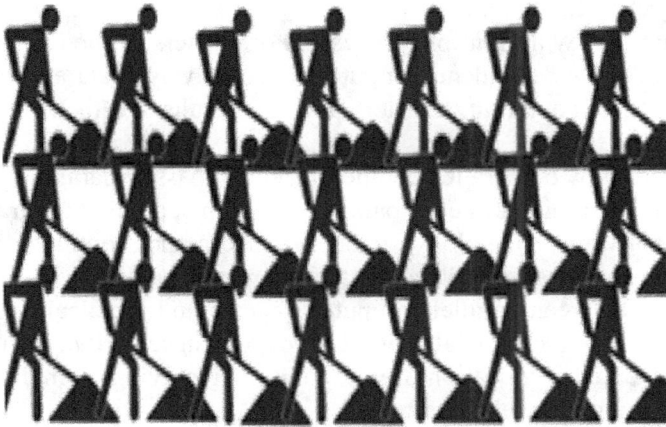

Fig. 5.2.   Task decomposition.

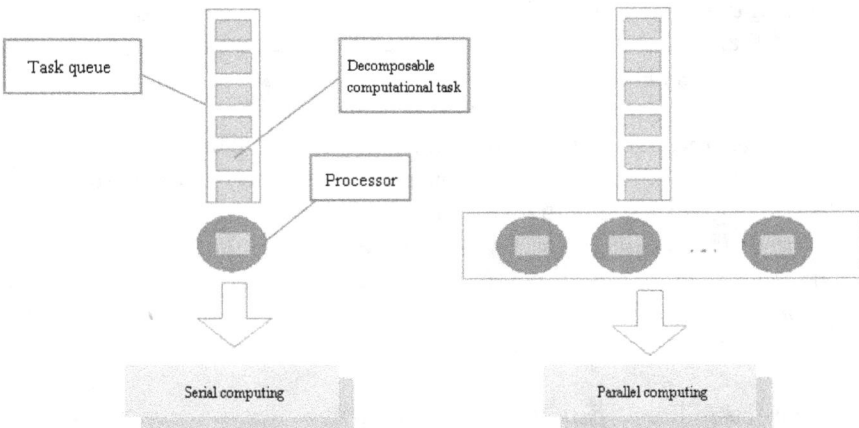

Fig. 5.3.   Differences between parallel and serial computing.

computer systems. This approach involves solving a large-scale computational problem by running small tasks on multiple components in cooperation. The basic idea is to use multiple processors to collaboratively solve the same problem, splitting the problem into several parts, each of which is independently computed by a different processor. In essence, a parallel computing system is a computer system composed of multiple computing units, characterized by high-speed computation, a large storage capacity, and high reliability. Parallel computing systems can be specially designed supercomputers with multiple processors or clusters formed by interconnecting several independent computers in some way. Data are processed by the parallel computing clusters and the results are then returned to the user.

The concept of parallel computing encompasses parallel computer architecture, compiler systems, parallel algorithms, parallel programming, parallel software technology, parallel performance optimization and evaluation, and parallel applications. Furthermore, parallel computing acts as a bridge between parallel computer systems and practical application problems. It provides crucial support for experts in scientific, engineering, and commercial application fields, aiding in solving domain-specific problems using parallel computers.

Therefore, three basic conditions are necessary for the successful implementation of parallel computing:

1. *Parallel computers*: These must include at least two or more processors that are interconnected and capable of communicating with each other over a network.
2. *Applications with parallelism*: Applications need to be decomposable into multiple sub-tasks which can be executed in parallel. The process of breaking down an application into several sub-tasks is known as the design of parallel algorithms.
3. *Parallel programming*: In the parallel programming environment provided by parallel computers, specific parallel algorithms are implemented and parallel programs are developed and run, achieving the goal of solving application problems in parallel.

The feasibility of parallel computing mainly lies in the fact that concurrency is a universal attribute of the physical world. Most computational problems with practical application backgrounds can be divided into multiple sub-tasks that can be computed in parallel.

5.1.2. *Levels of parallel computing*

Parallel granularity (Granularity) refers to the computational load performed between two parallel or interactive operations, which is essentially the size of the task. Parallel computing can be divided into different levels

Coarse

■ Program-level Parallelism

■ Subroutine-level Parallelism

■ Statement-level Parallelism

■ Operation-level Parallelism

■ Micro-operation-level Parallelism

Parallelism Granularity

Fine

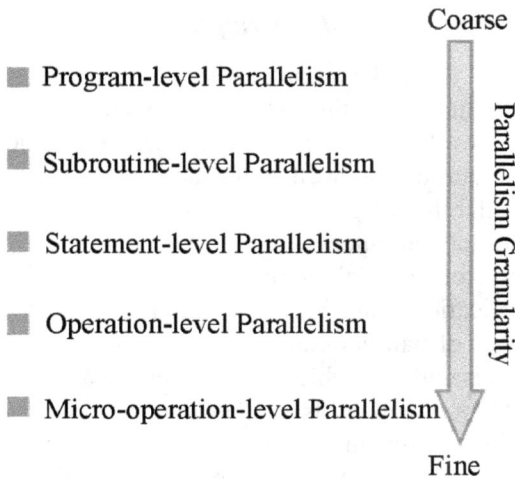

Fig. 5.4.   Hierarchical levels of parallel computing.

of granularity: program-level parallelism, subroutine-level parallelism, statement-level parallelism, operation-level parallelism, and micro-operation-level parallelism, as shown in Figure 5.4. The last three are mostly handled by hardware and compilers, while developers generally deal with the first two—program-level and subprogram-level parallelism. Typically, parallel computing focuses on the first two levels.

The degree of parallelism (DOP) refers to the number of processes executed simultaneously. The degree of parallelism is often inversely related to granularity: Increasing granularity decreases parallelism. However, increasing the degree of parallelism also increases system (synchronization) overhead.

In a data analysis task, if it can be divided into multiple independent computational tasks and assigned to different nodes for processing, this is known as program-level parallelism. Program-level parallelism is coarse-grained, and a problem that can achieve this level of parallelism is easy to execute in a cluster. Since the divided tasks are independent, the communication cost between sub-problems is very low, requiring minimal data transfer between cluster nodes. Each computational task at the program level can be considered independent, with no computational or data dependencies, and its parallelism is natural and macroscopic. Cloud computing and big data mainly focus on program-level parallelism.

5.1.3. *Development of parallel computers*

The emergence of parallel computers is a crucial precondition for the application of parallel computing. The development of parallel computers has been driven by the growing demand for large-scale scientific and engineering computations, as well as business processing and transaction handling. The scalability of problem-solving is one of the most important indicators of parallel computers.

The demand for parallel computing in large-scale scientific and engineering applications has been the main driving force behind the rapid development of parallel computers. The need for parallel computing in these areas is endless, which accelerates the advancement of parallel computing. Market demand is another major driving force for parallel computing. Various applications, such as weather forecasting, nuclear science, oil exploration, seismic data processing, and numerical simulations of aircraft, require computers capable of performing trillions or even quadrillions of floating-point operations per second. Parallel computing is a feasible way to meet these practical needs, further promoting the development of parallel computers. In addition to large-scale scientific and engineering applications, the rapid development of microelectronics and large-scale integrated circuits has also contributed to the advancement of parallel computers.

5.1.3.1. *Development of parallel computers over time*

The first parallel computer was developed in 1972 (ILLIAC IV, University of Illinois), consisting of 64 processors. It had good scalability but poor programmability. In 1976, the Cray-1 vector machine was introduced, dominating the supercomputing world for more than a decade. While Cray-1 was easy to program, its scalability was limited.

During the 1980s, parallel computers experienced rapid development, with early machines primarily using Multiple Instruction stream, Multiple Data stream (MIMD) parallel computers. The mid-1980s saw the emergence of shared-memory multiprocessors (SMPs), which gathered a group of processors in a single computer where multiple CPUs shared a memory subsystem and bus structure. This symmetrical multiprocessing significantly improved data processing capacity but had limitations in scalability and reliability, often encountering memory access bottlenecks. Later, more powerful parallel computers appeared, such as the Meiko (Sun) system with a 2D mesh connection, MIMD parallel computers connected via a hypercube, and the Cray Y-MP shared-memory vector multiprocessor, among others.

In the 1990s, parallel computing architectures became more unified, with Distributed Shared Memory (DSM), Massively Parallel Processing (MPP), and Cluster of Workstations (COW) being representative models. The focus during this period was on making each node increasingly independent, with the goal of turning each node into a fully functional workstation with its own hard drive and UNIX system. Nodes were connected by low-cost networks like Gigabit Ethernet. The Beowulf cluster became a typical example of COW, and the boundaries between COW and MPP began to blur.

Since 2000, parallel computing has made unprecedented leaps forward. Parallel computers built around COW models, using commercial off-the-shelf PCs in large-scale clusters, became the norm. Cluster architectures like Cluster, Constellation, and MPP are now based on clusters. In Cluster systems, each node contains multiple commercial processors, and nodes share memory internally. Nodes are connected by commercial cluster switches through a front-end bus, and storage is distributed across nodes. Each node runs Linux, along with GNU compilers and job management systems. The Constellation system consists of sub-parallel computers in each node, connected by commercial switches and utilizing distributed storage, with specialized operating systems, compilers, and job management systems running on each node. Figure 5.5 shows the development of parallel computers through the ages.

5.1.3.2. *Development of parallel computers by application characteristics*

The development of parallel computers can be broadly divided into the following two eras:

(1) *Specialized era*: This includes vector machines, MPP (Massively Parallel Processing) systems, SGI NUMA systems, and SUN large-scale SMP (Symmetrical Multi-Processor) systems, as well as China's ShenWei, Yinhe, and Sugon 1000 systems. The term "specialized" does not mean that these machines can only run specific applications, but rather that their components were specially designed. Their CPU

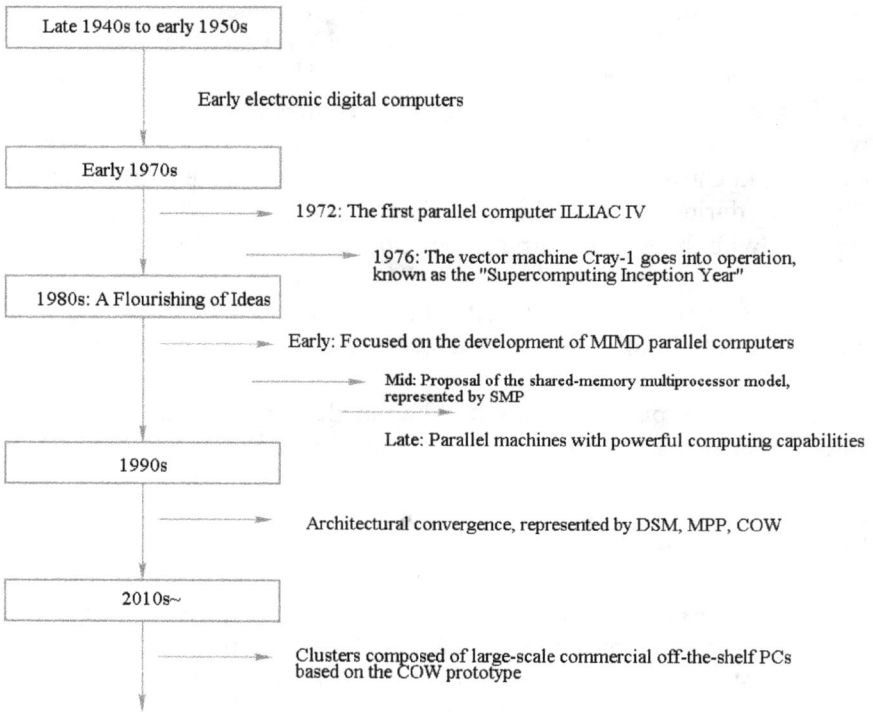

Fig. 5.5.   The evolution of parallel computing.

Note: COW, also known as a cluster, refers to a system where a group of computers runs the same software and is virtualized as a single host to provide services to clients and applications. This type of system has relatively low technical requirements, and users can even build a cluster themselves by connecting several servers or personal computers through Ethernet and using appropriate management and communication software. However, creating a high-performance, well-structured cluster with good Remote Access Service (RAS) capabilities is not an easy task. Almost all domestic and international computer manufacturers have their own cluster products, such as IBM's Sequoia and Mira, and the Tianhe series developed by the National University of Defense Technology, as well as Sugon's Nebula series.

boards, memory boards, I/O boards, and even operating systems were custom-made and cannot be used in other systems.

(2) *Popularization era*: As the price of high-performance computers dropped and the entry barriers for their use lowered, applications began to become more widespread. Two technological aspects played a crucial role in this. First, the commercialization trend meant that mass-produced commodity components became comparable in

performance to specialized high-performance computer components, and the standardization trend allowed these components to be integrated into a unified system. The development of the X86 processor, Ethernet, memory components, and the Linux operating system was decisive in this process. Second, one of the mainstream architectures of modern high-performance computers is the cluster system, whose technical foundation and industrial base have achieved industrial standardization, offering a very high cost-to-performance ratio.

Due to the high development and sale costs of vector machines and MPP systems, their market was somewhat limited. SMP systems, due to the limitations of their shared architecture, could not scale to very large systems. However, cluster systems, with their high cost–performance ratio, low investment risk, flexible structure, strong scalability, good general applicability, ability to inherit existing hardware and software resources, short development cycles, and programmability, became the dominant trend in parallel computer development.

The most basic hardware required for cloud computing is large server clusters connected in series. To solve the heat dissipation problem caused by densely connected servers, cloud computing data centers often use a "container-style" arrangement, where large server clusters are neatly placed in cabinet-like structures resembling containers. To ensure the efficiency of the cloud computing platform, large-scale server clusters must adopt serial connection technologies that feature scalability, data redundancy, fault tolerance, and load balancing. For example, Google's Atlanta and Dallas data centers serve as mutual backups, distributing computational tasks evenly across the server clusters to maintain load balance between them.

5.1.4. *Parallel computing and distributed computing*

Both parallel computing and distributed computing fall under the umbrella of High-Performance Computing (HPC), and their main purpose is to analyze and process large amounts of data. They both use parallelism to achieve higher computational performance by dividing large tasks into smaller ones. This similarity often leads to confusion between the two, but there are significant differences. Understanding their principles, characteristics, and appropriate application scenarios is crucial for a deeper comprehension of cloud computing.

First, let's briefly introduce the concept of distributed computing. Distributed computing mainly studies how computing is performed within distributed systems. A distributed system is a group of computers interconnected via a computer network. Distributed computing can place programs on the computer most suitable for running them, which is one of its core concepts, enabling resource sharing and load balancing.

Parallel computing is the opposite of serial computing. The main goal of parallel computing is to speed up problem-solving and scale up the problem size. Parallel computing emphasizes real-time performance and large-scale data processing. The tasks are often tightly interconnected, requiring synchronization between nodes. In other words, the tasks handled by parallel programs are highly interdependent, and every task block must be processed, as the results are interrelated and affect each other. Therefore, parallel computing requires that each computational result be absolutely correct.

Distributed computing, on the other hand, is the opposite of centralized computing. In distributed computing, tasks are relatively independent, meaning that if a task block's result has not been returned or is incorrect, it has little effect on the next task block. Distributed computing has lower real-time performance requirements and allows for computational errors (because multiple participants may compute each task, and the results are uploaded to a server for comparison, with significant differences verified). This means distributed computing does not emphasize synchronization, and tasks between nodes can be executed without communication or data transmission. There is no strict time limit for task execution between nodes.

A typical example of distributed computing is the Folding@home project, which analyzes the internal structure of proteins and related drug research. This large-scale project requires an enormous amount of computation that a single computer could never handle. Through distributed computing, the computational workload is divided into smaller chunks and assigned to many computers. Once the computation results are uploaded, they are combined to derive conclusions. Some consider distributed computing a special case of parallel computing.

In distributed computing, many task blocks can be skipped, meaning there is often a large amount of unnecessary data processing. Although distributed computing is fast, its true "efficiency" is relatively low. Distributed computing typically deals with "searching" problems, which are similar to brute-force methods. Every potential result from 0 to n is tested until the correct result is found. The answer might be found early or late in the search process, or it may never be found at all. In contrast, parallel computing deals with a finite number of task blocks, and it should be completed within a limited time frame.

Distributed computing programs are generally written in C++ or Java, without using MPI interfaces. Parallel computing programming, on the other hand, typically uses MPI or OpenMP.

5.1.5. *Parallel computing and cloud computing*

Cloud computing would not have emerged without support from other technologies, particularly parallel computing, big data technologies, and network technologies. Cloud computing is the inevitable result of advancements in these fields.

Cloud computing needs to address several challenges: transparent and elastic virtualization of computational resources, transparent and elastic virtualization of internal and external storage resources, data security guarantees, providing developers with a complete API, and ensuring a smooth transition for end users to cloud computing. Cloud computing abstracts everything into the cloud, so ordinary users no longer need to worry about where their data are stored, whether their applications need updates, or the dangers of computer viruses. All of these concerns are handled by cloud computing. The only thing regular users need to do is purchase the services they need from their favorite cloud service provider and pay accordingly. Cloud computing allows everyday users to enjoy high-performance computing because cloud computing centers can provide almost unlimited computational power. Elasticity in computing and storage is a defining feature of cloud computing.

The computing power provided by cloud computing is achieved through the parallelization of computers. By connecting multiple computers in parallel, faster computational speeds are attained, which is a simple yet effective method of achieving high-speed computing.

After the advent of large-scale parallel computers, cloud computing server clusters expanded significantly. With the number of servers in such clusters reaching tens of thousands or even more, cloud computing faces two major challenges: the high cost of system deployment and the frequent occurrence of node failures. In the cloud computing environment, computing and storage capacities adjust according to demand, meaning that cloud computing operates on an on-demand basis. This concept also applies to the requirements for servers in the cloud era, where the focus shifts from high-performance, fully configured servers to those that are simply "good enough." In cloud computing, clusters are typically built from large numbers of commercial off-the-shelf (COTS) PCs, which greatly reduce hardware costs. This cluster architecture is based on the parallel computing models established after 2000.

Given the large concentration of servers, server failures are common. Traditional architectures are sensitive to single points of failure, but cloud computing assumes single-point failures as a regular occurrence. Under this model, the failure of a single node does not affect the overall service provided by the system. Cloud computing systems are designed with these potential failures in mind, creating an infrastructure that accommodates untrusted nodes. In this server cluster model, server failures are accounted for in the system design, ensuring that they are hidden from developers and ordinary users. Data security is maintained through replication strategies, ensuring the integrity of data despite potential failures.

5.2. Cloud Computing Infrastructure—Cluster Technology

Cluster architecture is the mainstream architecture for high-performance computing today and is also dominant in the big data field. Cluster technology is a key technology supporting both cloud computing and big data systems.

5.2.1. *Basic concept of clustering*

A cluster is a collection of independent computers (nodes) connected through high-performance networks. Each node functions as a single computing resource for users and can also work collaboratively, presenting itself as a single, centralized computing resource for parallel computing tasks. A cluster is a cost-effective, easy-to-build, and scalable architecture. It has the following key characteristics:

(1) Each node in the cluster is a complete computer system, which can be a workstation, PC, or a symmetrical multiprocessor (SMP).
(2) The network connection typically uses commercial network devices such as Ethernet, FDDI, or optical fiber. Some commercial clusters also use dedicated network interconnections.

(3) The network interface is loosely coupled with the I/O bus of the nodes.
(4) Each node has a local disk.
(5) Each node runs its own independent operating system.

There are five critical issues to consider when designing a cluster system:

(1) *Availability*: A cluster system includes a middleware layer that provides availability services, such as checkpoints, failover, error recovery, and fault tolerance across all nodes. This allows the system to make full use of the redundant resources in the cluster and ensure long uptime for users.
(2) *Single system image (SSI)*: The difference between a cluster and a group of interconnected workstations is that the cluster can present itself as a single system. The cluster system also has an SSI middleware layer that combines the operating systems of each node to provide unified access to system resources.
(3) *Job management*: To achieve high system utilization, job management software in the cluster must provide functions like batch processing, load balancing, and parallel processing.
(4) *Parallel file system (PFS)*: Since many parallel applications in a cluster handle large amounts of data and perform numerous I/O operations, a high-performance parallel file system is essential to achieve optimal efficiency.
(5) *Efficient communication*: Cluster systems require a more efficient communication subsystem than MPP machines because cluster nodes are more complex, and the longer connections between nodes result in higher communication latency. Additionally, challenges like reliability, clock skew, and cross-talk arise.

5.2.2. *Classification of cluster systems*

An ideal cluster system hides the existence of multiple nodes from the user. To users, a cluster appears as a single computer system rather than multiple individual systems. Administrators can add or remove nodes in the cluster without the user being aware.

Cluster systems can be classified into the following four categories based on their functionality and structure:

(1) *High-availability cluster systems*: These systems achieve high availability by using backup nodes. If an active node fails, the backup node automatically takes over its workload. Such clusters are primarily used for mission-critical applications that require continuous service.

(2) *Load-balancing cluster systems*: In load-balancing clusters, all nodes participate in the workload, and the system distributes tasks evenly across nodes using scheduling algorithms like round-robin, least-loaded first, or consistent hashing.

(3) *High-performance cluster systems*: These clusters focus on providing powerful computational capabilities and are designed to complete complex tasks. High-performance cluster systems are commonly used in scientific computing, with many serving fields such as physics, biology, and chemistry.

(4) *Virtualized cluster systems*: With the widespread use of virtualization technology, server resources can be fully utilized and divided by creating multiple independent virtual machines on a single server. Management software is used to allocate and manage virtual resources. These clusters are referred to as virtual clusters, where the computing and storage resources usually reside on a single physical machine. Virtual cluster systems support cloud computing applications like virtual desktop technology.

Currently, cloud computing and big data systems based on cluster architectures often integrate multiple cluster types. These systems need to meet high availability requirements, achieve load balancing across nodes, and handle large data processing tasks. Big data systems like Hadoop and HPCC (High-Performance Computing Cluster) incorporate mechanisms from the first three cluster types, while cloud computing systems based on virtualization typically employ virtualized clusters.

5.2.3. *Cluster file systems*

The evolution of information recording methods has paralleled the development of human history, and the evolution of file systems has played a significant role in cloud computing. Data storage methods have a profound impact on the architecture of cloud computing systems. Traditional storage methods are generally based on centralized disk arrays, which are simple to set up and easy to use. However, since data are stored in a centralized location, data transmission across the network becomes inevitable during usage, placing significant pressure on the network.

With the advent of big data technologies, data-driven computation has become a problem that cloud computing systems need to address. Centralized storage faces considerable challenges, and the new concept of moving computation to the data has rendered centralized storage less effective. Cluster file systems emerged in response to these challenges. Currently, popular file systems like HDFS, GFS, and Lustre all belong to the category of cluster file systems.

Cluster file systems do not store data on a single node's storage device. Instead, they distribute data across different physical nodes' storage devices according to specific strategies. A cluster file system virtually integrates the storage space of every node in the system, forming a global logical directory. When accessing files, the logical directory maps to physical storage locations based on the internal storage strategy of the file system, allowing file locations to be determined. Compared to traditional file systems, cluster file systems are more complex. They must resolve issues like data consistency across different nodes and distributed locking mechanisms. Therefore, cluster file systems have become one of the core research topics in cloud computing technology.

Cluster file systems are categorized into different types. Based on storage access methods, they can be divided into shared storage cluster file systems and distributed cluster file systems. The former allows multiple computers to share the same storage space and jointly manage the files on it, also known as shared file systems. The latter lets each computer provide its own storage space, with each coordinating file management across all nodes. Examples of shared storage cluster file systems include Veritas CFS, Quantum Stornext, Blue Whale BWFS, and EMC MPFS. Distributed cluster file systems, commonly used in large-scale internet clusters, include HDFS, GFS, Gluster, Ceph, and Swift, known for their strong scalability, with some supporting up to 10,000 nodes.

Based on metadata management methods, cluster file systems can be classified into symmetric and asymmetric systems. In symmetric cluster file systems, all nodes play an equal role in managing file metadata, with high-speed synchronization and mutual exclusion operations between nodes. A typical example is Veritas CFS. In asymmetric cluster file systems, one or more nodes are dedicated to managing metadata, with other nodes frequently communicating with metadata nodes to obtain the latest metadata, such as directory listings and file attributes. Examples include HDFS, GFS, and BWFS. Cluster file systems can be configured as distributed + symmetric, distributed + asymmetric, shared + symmetric, or shared + asymmetric, with any combination being possible.

Based on file access methods, cluster file systems can be further divided into serial access and parallel access systems. Parallel access systems, also known as parallel file systems, allow clients to access data from any or multiple nodes in the cluster simultaneously, improving access speed. HDFS, GFS, and pNFS support parallel access but require specialized clients, whereas traditional NFS/CFS clients do not support parallel access.

5.3. Classification of Parallel Computing

Parallel computing can be classified into two types: temporal parallelism and spatial parallelism. Temporal parallelism refers to pipeline technology, which is a quasi-parallel processing technique that overlaps operations during the execution of multiple instructions in a program. Spatial parallelism refers to using multiple processors to execute computations concurrently. Parallel computing mainly focuses on spatial parallelism.

During the development of parallel computing, various technical methods and classification schemes have emerged. These include Flynn's classification based on instruction and data processing, classification by memory access structure, and classification based on application computational characteristics. The following is a brief introduction to these methods.

5.3.1. *Flynn's classification*

In 1972, a Stanford University professor, Michael J. Flynn, proposed a classic method for classifying computer architectures based on the most abstract aspects of instruction and data processing, known as Flynn's classification. Flynn's classification focuses on instruction streams, data streams, and multiplicity. According to Flynn's classification, spatial parallel computing can be divided into two types: Single Instruction stream, Multiple Data stream (SIMD), Multiple Instruction stream, Multiple Data stream (MIMD), as shown in Figure 5.6.

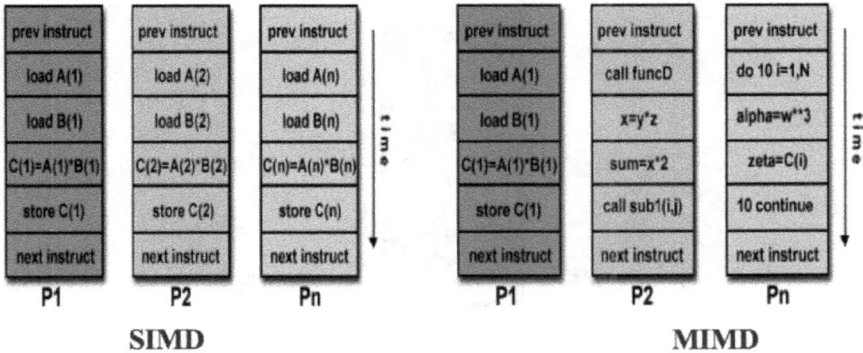

prev instruct	prev instruct	prev instruct
load A(1)	load A(2)	load A(n)
load B(1)	load B(2)	load B(n)
C(1)=A(1)*B(1)	C(2)=A(2)*B(2)	C(n)=A(n)*B(n)
store C(1)	store C(2)	store C(n)
next instruct	next instruct	next instruct
P1	**P2**	**Pn**

SIMD

prev instruct	prev instruct	prev instruct
load A(1)	call funcD	do 10 i=1,N
load B(1)	x=y*z	alpha=w**3
C(1)=A(1)*B(1)	sum=x*2	zeta=C(i)
store C(1)	call sub1(i,j)	10 continue
next instruct	next instruct	next instruct
P1	**P2**	**Pn**

MIMD

Fig. 5.6.   Parallel computing classified by Flynn's taxonomy.

(1) *Single instruction stream, multiple data stream (SIMD)*: This technique uses one controller to manage multiple processors, executing the same operation on each piece of data in a dataset (also known as a "data vector") concurrently, thus achieving spatial parallelism. SIMD achieves data-level parallelism, and its typical representatives include vector processors and array processors.

The key to SIMD technology is executing multiple operations simultaneously within a single instruction to increase processor throughput. To achieve this, SIMD CPUs have multiple execution units, all controlled by a single instruction unit. The central controller sends instructions to each processing unit, and the entire system requires only one central controller and one copy of the program, with all computations synchronized. Modern single-core computers generally belong to the SIMD category.

(2) *Multiple instruction stream, multiple data stream (MIMD)*: In MIMD systems, different processors can execute different instructions on different pieces of data during any clock cycle, meaning multiple instruction streams operate simultaneously on separate data streams. MIMD uses multiple controllers to asynchronously control multiple processors, achieving comprehensive parallelism across jobs, tasks, instructions, and arrays. The latest multi-core platforms fall under the MIMD category, such as Intel and AMD dual-core processors.

Serial machines commonly use Single Instruction stream, Single Data stream (SISD), as shown in Figure 5.7. An SISD system has only one central processing unit, as seen in workstations and single

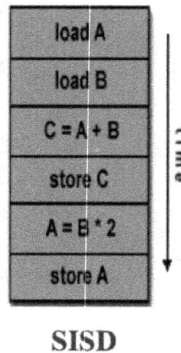

SISD

Fig. 5.7.   Single instruction data flow.

computing servers, and it is not a parallel system. SISD is a traditional sequential processing system where the instruction unit decodes one instruction at a time and assigns data to one operational unit. Single-processor systems using pipeline techniques are sometimes also classified as SISD. SISD hardware does not support any form of parallel computing, and all instructions are executed serially. Early computers, such as IBM PCs, were SISD machines.

5.3.2. Classification by computational characteristics of applications

Based on the computational characteristics of applications, parallel computing can be classified into the following two categories:

(1) *Data-intensive parallel computing*: This type of parallel computing uses data parallel methods to handle large datasets (typically in the TB or PB range). Data-intensive parallel computing is used for I/O-bound applications or those that need to process massive amounts of data, with most processing time spent on I/O, data movement, and data processing. Parallel processing of data-intensive applications usually involves partitioning or subdividing the data into multiple parts that can be processed independently in parallel on appropriate computing platforms, with the results combined to produce a complete output. The larger the dataset, the greater the benefits of parallel data processing. Data-intensive processing requirements are usually scaled linearly according to the size of the total amount of data, and are very suitable for direct parallel processing.

(2) *Computation-intensive parallel computing*: Computation-intensive applications are those where the majority of execution time is spent on computation rather than I/O, usually requiring small amounts of data. Parallel processing of computation-intensive applications often involves parallelizing algorithms within the application processes and breaking the overall process into individual tasks, which can be executed in parallel on appropriate computing platforms to achieve higher performance than serial processing. In computation-intensive applications, multiple operations are executed simultaneously, with each operation addressing a specific part of the problem, a concept often referred to as task parallelism. Traditional high-performance computing applications, such as weather forecasting, high-resolution nuclear simulations, and scientific computing for image processing, are largely computation-intensive.

5.3.3. *Classification by structural models*

Parallel computing can also be classified by structural models into the following categories: Parallel Vector Processor (PVP), Symmetrical Multi-Processor (SMP), Distributed Shared Memory (DSM), Massively Parallel Processor (MPP), and Cluster of Workstations (COW), as shown in Figure 5.8.

5.3.3.1. *Parallel vector processor (PVP)*

PVP is a type of parallel architecture model consisting of a small number of highly powerful custom vector processors. It uses high-bandwidth crossbar switches and high-speed data access models. PVP typically does not use a cache but instead employs a large number of vector registers and instruction caches, making programming for this model highly demanding.

5.3.3.2. *Symmetrical multi-processor (SMP)*

SMP systems use shared memory, where any processor can directly access any memory address, with equal access latency, bandwidth, and probability across the system. The system is symmetrical. Microprocessors in SMP systems are typically fewer than 64, as buses and crossbar switches become difficult to expand once built. Typical examples include IBM R50, SGI Power Challenge, SUN Enterprise, and Sugon No. 1.

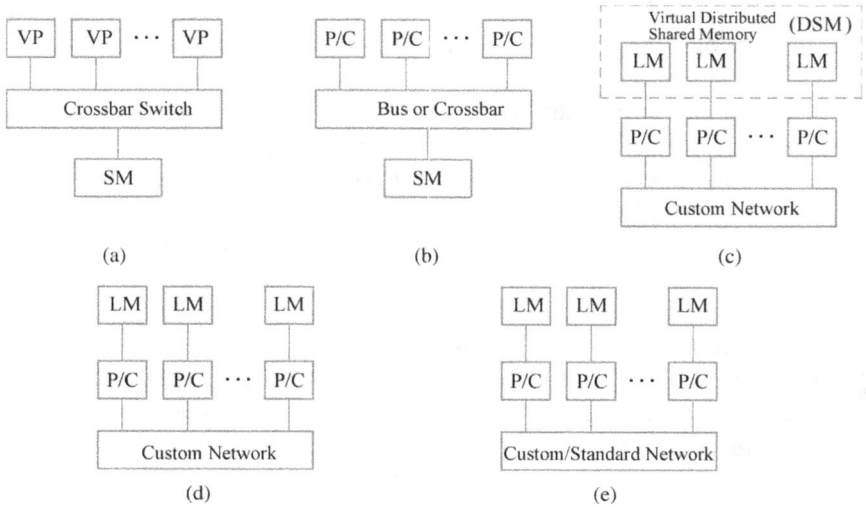

Fig. 5.8.   Classification of parallel computing models by structure. (a) PVP, (b) SMP, physically a single address space, (c) DSM, logically a single address space, (d) MPP, physically and logically multiple address spaces, and (e) Cluster/COW, physically and logically multiple address spaces.

5.3.3.3. *Distributed shared memory (DSM)*

DSM is also a shared-memory system, logically (from the user's perspective) providing shared memory, though memory modules are physically distributed across processors. This structure is also called Cache-Coherent Non-Uniform Memory Access (CC-NUMA). DSM microprocessors typically range from 16 to 128, with examples like SGI Origin 2000 and Cray T3D.

DSM (Distributed Shared Memory) is organized by nodes, where each node has one or more CPUs. It uses a specialized high-performance interconnection network and distributed storage, meaning the memory modules are distributed across the nodes. DSM employs a single memory address space, where all memory modules are uniformly addressed by hardware. Each node can access both its local memory units and the memory units of other nodes. DSM runs a single operating system and can scale to hundreds of nodes.

The main difference between DSM and SMP is that DSM has physically distributed local memory across nodes, but logically forms a shared memory.

5.3.3.4. *Massively parallel processor* (*MPP*)

A Massively Parallel Processor (MPP) uses distributed memory both physically and logically, allowing it to scale to hundreds or even thousands of processors (either microprocessors or vector processors). MPP systems use high-bandwidth, low-latency interconnection networks, which require specialized design and customization. Typical examples include CRAY T3E (2048 processors), ASCI Red (3072 processors), IBM SP2, and Sugon 1000.

In MPP systems, each node is relatively independent and has one or more microprocessors. Each node has its own operating system and independent memory, avoiding memory access bottlenecks, though each node can only access its own memory module. As a result, MPP systems offer excellent scalability.

MPP systems follow an asynchronous MIMD (Multiple Instruction, Multiple Data) architecture, where the program system consists of multiple processes, each with its own private address space. Communication between processes is achieved through message passing.

5.3.3.5. *Cluster of workstations* (*COW*)

Also known as a Network of Workstations (NOW), COW systems consist of complete workstations, each with their own hard drive and operating system (e.g., UNIX). Nodes are interconnected through high-performance or low-cost networks like Gigabit Ethernet, and message-passing software enables communication and load balancing. Typical examples include Beowulf clusters and systems like Sugon 3000 and 4000, and ASCI Blue Mountain. The distinction between COW and MPP has become increasingly blurred over time.

5.4. Technologies Related to Parallel Computing

The main idea of parallel computing is to decompose a complex problem into several parts and assign each part to an independent processor (computing resource) for calculation. For different problems, parallel computing requires special parallel architectures, independent algorithm design, and special programming models. To make parallel computing feasible, a variety of technologies need to be used to design and develop the operating environment of parallel computing.

5.4.1. *Key technologies of parallel computing*

The basic conditions of parallel computing include hardware (parallel computer), parallel algorithm design, and parallel programming environment. At present, the key technologies of parallel computing mainly include four parts: architecture, algorithm design and analysis, implementation technology, and application, as shown in Figure 5.9.

5.4.1.1. *Architecture of parallel computer*

Parallel computers focus on two key points: The first involves the components, that is, the hardware. The components of the parallel computer architecture include nodes, interconnect networks, and memory. A node can be composed of one or more processors; the interconnect network refers to connecting nodes to form a network environment; memory refers to multiple groups of storage modules. The second involves the structural model. Typical structural models of parallel computers include PVP, SMP, DSM, MPP, and COW. These structural models have been introduced in Section 5.3.3 and will not be repeated here. The parallel computer architecture includes high-end high-performance computers and low-end popular computers.

5.4.1.2. *Design and analysis of parallel algorithm*

The design and analysis of parallel algorithm includes the design and analysis of parallel algorithm, algorithm library and test library.

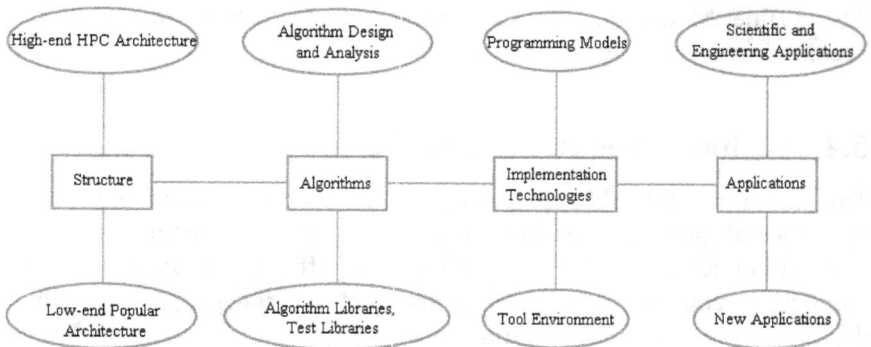

Fig. 5.9.   Key technologies of parallel computer.

The main idea of parallelization is to divide and conquer. The design ideas of parallel algorithms can start from two perspectives:

(1) *Domain decomposition*: According to the method of processing data, multiple relatively independent data areas are formed, which are processed by different processors to achieve data parallelism.
(2) *Task or function decomposition*: According to the solution process of the initial problem, the task is divided into several sub-tasks, and each sub-task completes a part of the entire work to achieve task-level parallelism or functional parallelism. This process should focus on the completed calculation rather than the calculation of the operating data. The process of task or function decomposition is shown in Figure 5.10.

Parallel algorithms are algorithms suitable for implementation on parallel computers. Good parallel algorithms should give full play to the potential performance of parallel computers. The parallel program design model (Parallel Program Model) is a collection of program abstractions and a concept built on the level of hardware and memory architecture. Commonly used program design models include the message-passing model, shared variable (storage) model, and data parallel model.

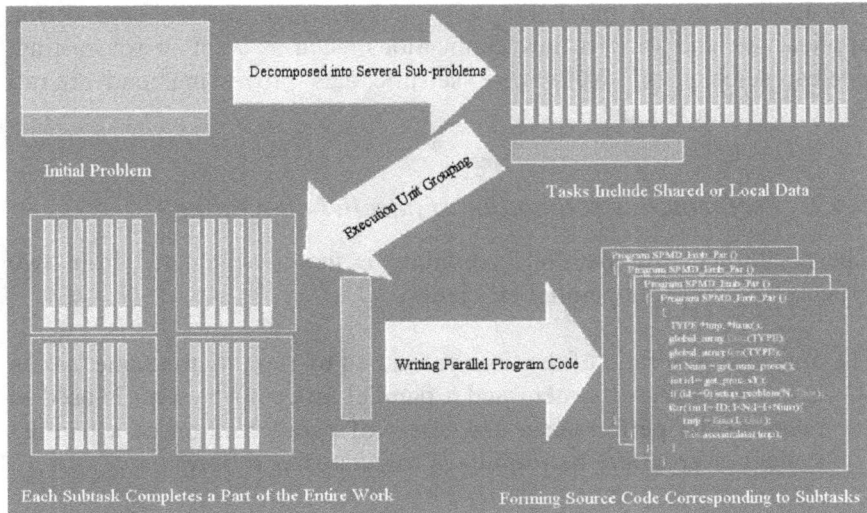

Fig. 5.10. Process of task or function decomposition.

5.4.1.3. *Parallel implementation technology*

Parallel implementation technology can be divided into three categories, namely, thread library, message-passing library, and compiler support. Thread libraries (such as POSIX* threads and Windows* API threads) can achieve explicit control over threads; if fine-grained management of threads is required, these explicit thread technologies can be considered. With the help of message-passing libraries (such as the Message Passing Interface, MPI), applications can simultaneously utilize multiple computers without having to share the same memory space.

Parallel implementation technology includes programming implementation and performance optimization, and it contains parallel programming models and environment tools for parallel programming.

The parallel programming environment mainly includes the operating system and programming language. The mainstream operating systems for parallel computers are UNIX/Linux. For example, IBM's AIX, HP's HPUX, Sun's Solaris, and SGI's IRIX are all variants of UNIX. The programming languages for parallel computing are Fortran 77/90/95 and C/C++.

5.4.1.4. *Application of parallel computing*

The application of parallel computing includes scientific and engineering applications and various new applications, such as weather forecasting, nuclear science, oil exploration, seismic data processing, and aircraft numerical simulation.

5.4.2. *Performance estimation of parallel computing*

The performance of parallel computing is mainly evaluated from two aspects: speedup and parallel efficiency.

(1) *Speedup*: The ratio of the time consumed by running the same task in a single-processor system and a parallel-processor system is used to measure the performance and effect of parallelization of a parallel system or program. Its formula is calculated as follows:

$$\text{Speedup} = T1/Tn,$$

where speedup is the speedup ratio, $T1$ is the running time under a single processor, and Tn is the running time in a parallel system with n processors.

(2) *Parallel efficiency*: This is the ratio of the speedup to the number of processors used. Parallel efficiency represents the execution efficiency of each processor on average when a parallel algorithm is executed on a parallel machine, as shown in Figure 5.11.

As can be seen from the left side of Figure 5.11 (Parallel Portion of Code, the parallel part of the code), the speedup ratio gradually increases with the percentage of the parallel code. Only when the parallel program code reaches more than 95% is the speedup ratio relatively ideal, and its parallel efficiency can be fully exerted. If the parallel code of the program is only 25%, its parallel efficiency is basically equal to zero, and even if there are more processors, it is useless. As can be seen from the right side of Figure 5.11, there are two important points. First, when the parallel program code reaches 95% and the parallel program code reaches 90%, the speedup ratio is very different, so their parallel efficiencies will be several orders of magnitude different. Second, when the number of processors reaches 1000 nodes, even if more processors are added later, the speedup ratio will no longer change; that is, its parallel efficiency will no longer improve.

From these points, it can be seen that it is not easy to write a completely heterogeneous (MIMD) parallel program for a computer with 1000 processors.

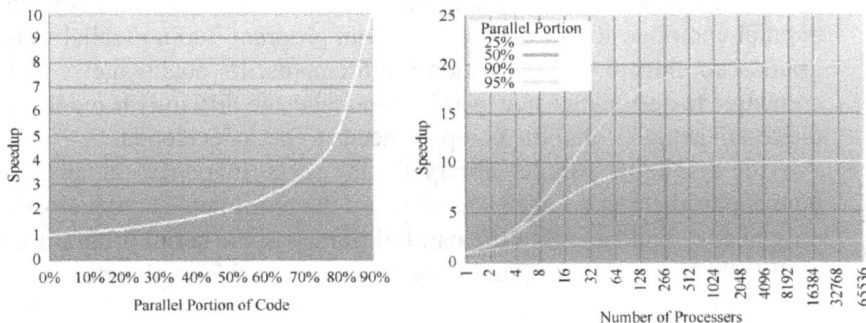

Fig. 5.11. Parallel efficiency diagram.

5.5. Parallel Program Design—MPI Programming

At present, the representative technology of parallel computing is the Message Passing Interface (MPI). MPI proposes a function interface description based on message passing, but MPI itself is not a specific implementation, only a standard description. We can find many shadows of MPI in Hadoop. For example, the file system and Map/Reduce processing in Hadoop are master–slave structures, and the master–slave structure is an important design method of MPI parallel programs. Therefore, understanding the MPI programming method in the era of parallel computing is beneficial to understanding some technical foundations and concepts in cloud computing. Therefore, the core technologies of parallel program design using MPI will be introduced later, so that readers can enter cloud computing from parallel computing.

5.5.1. *Introduction to MPI*

There are mainly three method of parallel program design:

(1) Design a new parallel language. Its advantage is that the implementation of parallel programs is simple and convenient; the disadvantage is that there is no unified standard, and the difficulty and workload of designing the language are significant.

(2) Expand the syntax of the serial language to support parallel features. The idea is to use the parallel expansion part of the serial language as the comment (annotation) of the original serial language. For the serial compiler, the parallel expansion part will have no effect, and for the parallel compiler, it will convert the serial program into a parallel program according to the annotation requirements. Its advantage is that compared to designing a new parallel language, the difficulty is reduced; the disadvantage is that the compiler needs to be redeveloped.

(3) Provide a callable parallel library for the serial language. Its advantage is that there is no need to redevelop the compiler. The developer only needs to add a call to the parallel library in the serial program to realize the design of the parallel program. From this explanation, the greater the change to the serial language, the greater the implementation difficulty. The relationship between the implementation method of parallel language and the implementation difficulty is shown in Figure 5.12.

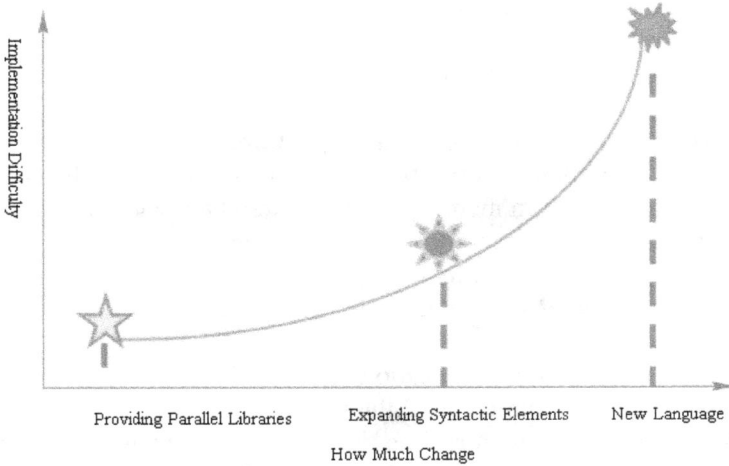

Fig. 5.12. Relationship between the implementation method of parallel language and the implementation difficulty.

At present, the most commonly used methods are to provide parallel libraries and expand syntax components to carry out parallel program design. The Message Passing Interface (MPI) implements the programming paradigm of distributed memory architecture by providing standardized parallel programming library functions. It is the standard specification of the message-passing function library. MPI is a new library description, not a language. MPI is a representative of a standard or specification, not a specific implementation of it. MPI is a message passing programming model, and become the representative and standard of this programming model. It was developed by the MPI Forum and supports Fortran, C, and C++.

Message-passing parallel program design requires users to realize data exchange between processors by explicitly sending and receiving messages. Each parallel process has its own independent address space, and mutual access cannot be directly carried out and must be realized through explicit message passing. This programming method is the main programming method adopted by massively parallel processors (MPPs) and clusters. Since message-passing program design requires users to decompose problems well and organize data exchange between different processes, the parallel computing granularity is large and is particularly suitable for large-scale scalable parallel algorithms. At present, message passing is a very important parallel program design method in the field of parallel computing.

The goals of designing MPI are as follows:

- Provide an application programming interface.
- Improve communication efficiency, including avoiding multiple repeated copies from memory to memory and allowing the overlap of computation and communication.
- Provide implementation in a heterogeneous environment.
- The provided interface can facilitate the invocation of C language and FORTRAN 77.
- Provide a reliable communication interface. That is, users do not need to deal with communication failures.
- The interface design should be thread-safe (allowing an interface to be called by multiple threads simultaneously).
- The semantics of the interface are language-independent.

MPI has the following characteristics:

- Communication mechanism based on message passing.
- Combines the parallel library of the serial language.
- Supports common serial languages such as Fortran, C, and C++.
- Good portability.
- Flexible and simple programming method. MPI provides a standard for writing message-passing programs, which is independent of language and platform and can be widely used. Using it to write message passing program is not only practical, portable, efficient and flexible, but also does not change much with the existing implementation (that is, most of the current MPI implementations provide the main functions of MPI).

At present, the main implementations of MPI are shown in Table 5.1.

5.5.2. *Implementation of a simple MPI program*

When writing an MPI program, two questions usually need to be answered: First, how many processes are used for parallel computing of the task? Second, it is necessary to know which process is running on each node?

Table 5.1. Main implementations of MPI.

Realization Name	Manufacturing Unit	Sites
MPICH	Argonne & MSU	http://www.mpich.org/
OpenMPI	Universities, research institutes, and large corporations	http://www.open-mpi.org/
LAM	Ohio State University	http://www.lam-mpi.org/

```
#include "mpi.h"        /*Import the header file containing the MPI library functions*/
#include <stdio.h>      /*Import the C language header file*/
/**
 * Using C language as the host language, call the MPI library to write a simple Hello.c parallel program
 */
int main(int argc, char *argv[])
{
    MPI_Init(&argc, &argv);    /**MPI Library Function/
    printf("hello world!\n");  /*C Library Function*/
    MPI_Finalize();            /*MPI Library Function*/
    return 0;
}
```

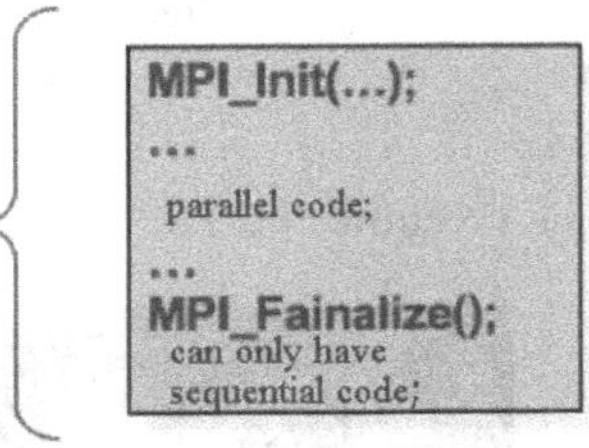

```
MPI_Init(...);
...
parallel code;
...
MPI_Fainalize();
can only have
sequential code;
```

Fig. 5.13. Screenshot of program code.

First, the simplest MPI parallel program in the form of C language is given. The task of this program is to print the words "Hello World!" on the terminal. The program code file Hello.c is shown in Figure 5.13.

Before writing the program, it is necessary to be clear about the C and MPI function conventions: The mpi.h header file must be included; the MPI function returns an error code or the MPI_SUCCESS success flag; the MPI_prefix and only the first letter after MPI_ and MPI_ are capitalized, and the rest are lowercase.

Then compile and run "Hello World!".

(1) Start 3 machine nodes in the machine cluster.

```
[mpi@node1 ~]$ mpdboot -n 3
```

(2) Compile the Hello.c program to generate the Hello.o program.

```
[mpi@node1 test3]$ mpicc -o hello.o hello.c
```

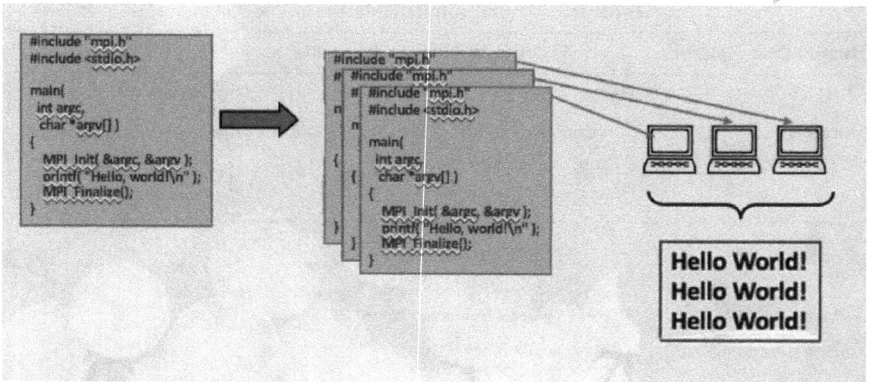

Fig. 5.14.   Execution process of "Hello World!".

(3) Execute the Hello.o program.

```
[mpi@node1 ~]$ mpirun -np 3 test3/hello.o
hello world!
hello world!
hello world!
```

Note: The executable program Hello.o must be in the test3 directory of the three machine nodes at the same time.

"Hello World!" is in SPMD format, and its execution process is shown in Figure 5.14. The code between MPI_init() and MPI_Finalize() is parallel code, which will be executed on all nodes. Since 3 nodes are deployed in this parallel environment, this parallel code will run once on three nodes, resulting in three Hello worlds! Note that MPI_init() and MPI_Finalize() are the start function and end function of the parallel program, respectively.

5.5.3. *MPI messages*

In the MPI program, all the information transmitted by inter-process communication is called a message. Through the transmission of messages, all the machine nodes in the entire cluster are controlled to work together, the underlying heterogeneous system is abstracted, and the program's portability is realized. A message consists of a message envelope and the message content:

(1) The message envelope indicates the object and related information for sending or receiving the message. The message envelope consists of the source/destination, tag, and communication domain. The format of the message envelope is similar to the cover of an envelope, as shown in Figure 5.15.

The destination is the recipient of the message and is determined by the parameters of the send function. The source is the provider of the message, which is implicitly determined by the sending process and uniquely identified by the rank value of the process. If the source/destination is implicit, it is group communication.

The tag is used to distinguish different messages of the same process so that the program can process the arriving messages in an orderly manner. It is necessary but not sufficient because the choice of "tag" has a certain degree of randomness. MPI uses a new concept "context" to expand the "tag". It is allocated during system runtime and is not allowed to be configured uniformly. If the "tag" is implicit, it is group communication, and the matching of messages is determined by the sequence of communication statements.

(2) The message content (data) indicates the entity data part that this message will transmit. The message content consists of the starting address, the number of data, and the data type, as shown in Figure 5.16.

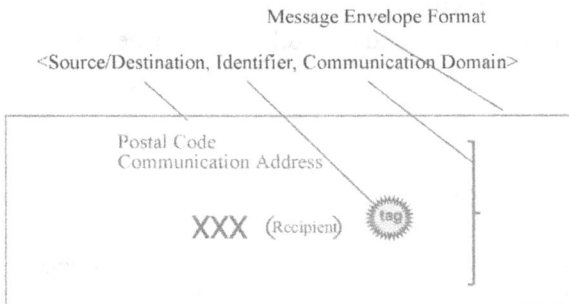

Message Envelope Format

<Source/Destination, Identifier, Communication Domain>

Postal Code
Communication Address

XXX (Recipient) (tag)

Fig. 5.15.   Format of message envelope.

Content (data) has variable forms in group communication.
<starting address, number of data, data type>

Six

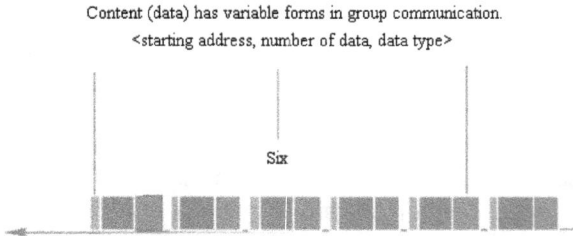

Fig. 5.16.   Organization form of message content.

5.5.4. *Message-passing process of MPI*

The message-passing process of MPI is divided into three stages: message assembly, message sending, and message disassembly, as shown in Figure 5.17.

The principles of message-passing parallel program design are as follows:

(1) Users must realize data exchange between processors by explicitly sending and receiving messages.
(2) Each parallel process has its own independent address space, and mutual access cannot be directly carried out and must be realized through explicit message passing.
(3) This programming method is the main programming method adopted by massively parallel processors (MPPs) and clusters.

Since message-passing program design requires users to decompose problems well and organize data exchange between different processes,

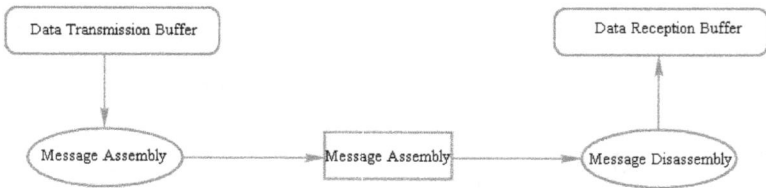

Fig. 5.17.   Message-passing process of MPI.

the parallel computing granularity is large, which is especially suitable for large-scale scalable parallel algorithms. Message passing is a very important parallel program design method in the current field of parallel computing.

5.5.5. *Commonly used basic functions of MPI*

Although the total number of MPI call interfaces is large, according to the actual experience of writing MPI, the number of commonly used MPI functions is limited. The following introduces 6 of the most basic MPI functions:

(1) *MPI_Init*: Start the MPI environment and mark the beginning of the parallel code. The function prototype is as follows:

```
int MPI_Init(int *argc, char **argv)
```

   It is usually the first MPI function to be called. Except for MPI_Initialized (to test whether MPI_Init has been executed), all other MPI functions should be called after it. MPI obtains the command-line parameters through argc and argv.

(2) *MPI_Comm_size*: Obtain the number of processes included in the group specified in the communication space comm. The function prototype is as follows:

```
int MPI_Comm_size(MPI_Comm comm, int *size)
```

(3) *MPI_Comm_rank*: Obtain the rank value of this process in the communication space comm, that is, the logical number in the group (starting from 0, similar to the process ID). The function prototype is as follows:

```
int MPI_Comm_rank(MPI_Comm comm, int *rank)
```

(4) *MPI_Send*: Standard blocking send message; the function prototype is as follows:

```
int MPI_Send(void *buff, int count, MPI_Datatype
    datatype, int dest, int tag, MPI_Comm comm)
```

where buff is the message-sending buffer; count is the number of messages of the specified data type MPI_Datatype, not the number of bytes; dest is the destination of the sent message; tag is the message tag; and comm is the communication space or communication domain.

(5) *MPI_Recv*: Standard blocking receive message; the function prototype is as follows:

```
int MPI_Recv(void *buff, int count, MPI_Datatype
    datatype, int source, int tag, MPI_Comm comm, MPI_
    Status *status)
```

where buff is the message-receiving buffer; count is the number of messages of the specified data type MPI_Datatype, not the number of bytes; source is the source of the sent message; Tag is the message tag; comm is the communication space or communication domain; and status is to record the message-receiving status (success or failure).

(6) *MPI_Finalize*: Mark the end of the parallel code and end other processes except the main process. The function prototype is as follows:

```
int MPI_Finalize(void)
```

Its function is to exit the MPI environment, and all processes must be called to exit normally. However, the serial code can still run on the main process (rank = 0), but there can be no more MPI functions (including MPI_Init).

5.5.6. *Parallel program with message passing*

Through the content of the previous sections, we have clearly understood that in the MPI program, inter-process communication is realized by passing messages. To facilitate understanding of message passing implementation in MPI, the following demonstration program utilizes C as the host language in conjunction with MPI's parallel libraries to enable parallel execution across three computing nodes. This MPI program is the Hello program with message passing.

5.5.6.1. *HelloWord.c program code*

The program code is as follows:

```
#include "mpi.h"
main(int argc, char* argv[])
{
int p; /Number of processes, this variable is a
    variable with the same name in each processor/
int my_rank; /Process ID, storage is distributed/
MPI_Status status; /Message receiving status variable,
    storage is distributed/
char message[100]; /Message buffer, storage is
    distributed/
MPI_Init(&argc, &argv); /Initialize MPI/
/This function is called once by each process to
    obtain its own process rank value/
MPI_Comm_rank(MPI_COMM_WORLD, &my_rank);
/This function is called once by each process to
    obtain the number of processes/
MPI_Comm_size(MPI_COMM_WORLD, &p);
if (my_rank!= 0)
{ /Create a message/
sprintf(message, "Hello Word, I am %d!", my_rank);
/The sending length is taken as strlen(message)+1 so
    that \0 is also sent out together/
MPI_Send(message,strlen(message)+1, MPI_CHAR, 0,99,
    MPI_COMM_WORLD);
}
else
{ /*my_rank == 0 */
for (source = 1; source <= 2; source++)
{ /Specify a parallel environment for 3 processes/
MPI_Recv(message, 100, MPI_CHAR, source, 99, MPI_
    COMM_WORLD, &status);
printf("%s\n", message);
}
}
MPI_Finalize(); /Close MPI, marking the end of the
    parallel code segment/
} /main/
```

5.5.6.2. *Compilation, running, and analysis of HelloWord.c*

(1) Compilation

Execute the command to generate the executable code of a.out by default: mpicc HelloWord.c.

The above command mpicc realizes the automatic generation of the default a.out file. If you want to generate your own named file, such as the generated file named HelloWord, you can execute the code command: mpicc-o HelloWord HelloWord.c.

(2) Running

When the first step of compilation is completed, the file can be executed, and the command is as follows: mpirun –n 3./a.out or mpirun –n 3./HelloWord.

Among them, 3 is the number of machine nodes specified to execute the parallel program, which is specified by the user. a.out or HelloWord is the MPI parallel program to be run. The case execution process is shown in Figure 5.18.

After the above-mentioned program is executed, the following results will be printed and output on the No. 0 machine node:

```
HelloWord, I am 1
HelloWord, I am 2
```

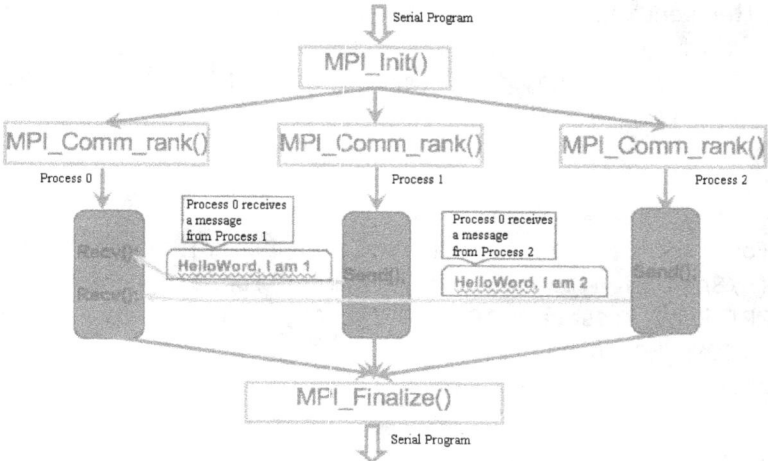

Fig. 5.18.   Case execution process.

From the above two examples, it can be seen that the parallel program is written in C language, and the execution of the parallel program can be completed by calling the MPI function library. When calling the parallel program of the MPI function library, the following conventions are followed:

- The program must include the mpi.h header file at the beginning.
- The MPI function returns an error code or the MPI_SUCCESS success flag.
- MPI_prefix and only the first letter after the MPI name and MPI_ are capitalized, and the rest are lowercase.
- The parallel part of the program starts with MPI_Init() and ends with MPI_Finalize(), and the other parts are the serial parts of the program.

Exercises

1. Explain what parallel computing is. Briefly summarize the development of parallel computing.
2. Briefly explain the concept of a cluster and the classification of cluster systems.
3. What are the classifications of parallel computing?
4. Briefly explain the four design models of parallel computing.
5. What are the main methods of parallel program design? Explain each of them.

Chapter 6

OpenStack: The Powerful Platform of IaaS

Currently, OpenStack has become the mainstream platform for IaaS. This chapter provides an in-depth introduction to OpenStack, detailing its design framework as well as the design of some of its key components.

6.1. OpenStack Architecture

OpenStack is currently the most popular IaaS platform, featuring an extremely large and rich set of modules and interfaces that support a wide variety of functions and services. OpenStack adopts a modular architectural approach, designed with the core objectives of scalability and stability of core service functions. It is this modular design that allows each component of OpenStack to be deployed independently, offering great flexibility. However, installing and using OpenStack can be challenging without an understanding of its internal structure and relationships. Therefore, understanding the role of each major module and their interdependencies is greatly beneficial for learning and utilizing the OpenStack platform.

Logically, the OpenStack architecture is mainly divided into three parts: control, compute, and networking. The control part provides Application Programming Interface (API) services, database services, and a message bus; the compute part offers virtual machine (VM) layer services that control the operation of VMs; the networking part is responsible for integrating physical and virtual network resources, providing network interconnection services for the entire OpenStack platform, including the

167

networks used by VM instances. Other services in OpenStack are responsible for collaborating to complete virtualization tasks, such as storage, authentication, and metering.

As shown in Figure 6.1, a simplified framework diagram of OpenStack is presented. From the diagram, it can be seen that OpenStack is composed of several major modules: DASHBOARD (Dashboard Service Module), COMPUTE (Compute Service Module), BLOCK STORAGE (Block Storage Service Module), NETWORKING (Networking Service Module), IMAGE SERVICE (Image Service Module), OBJECT STORAGE (Object Storage Service Module), and IDENTITY SERVICE (Identity Service Module). Readers should note that OpenStack is not only made up of these modules; there are other components as well, such as monitoring and orchestration.

The following provides a brief introduction to each of the main modules of OpenStack:

(1) *Dashboard*: The Dashboard service module, whose project name is Horizon, provides a web-based management interface for users. It allows for convenient and intuitive management of the OpenStack platform. Of course, if the user is advanced and prefers to manage the OpenStack platform using only the command line interface, this module can be omitted from installation.

(2) *Compute*: The Compute service module, with the project name Nova, is an indispensable core component of OpenStack. It provides a computing platform for users and is primarily responsible for interfacing with virtualization platforms such as KVM and Xen.

(3) *Block Storage*: The Block Storage service module, with the project name Cinder, is designed to integrate all storage resources within the

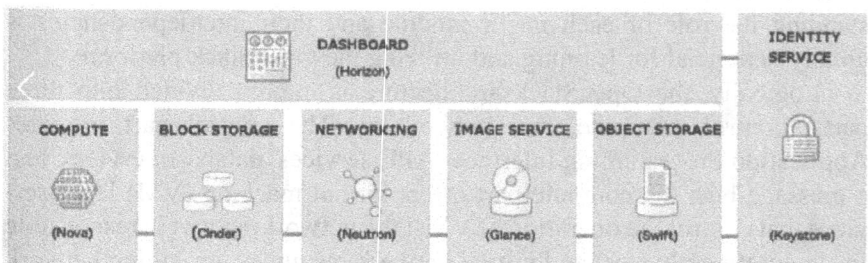

Fig. 6.1.   A simple architecture figure of OpenStack.

OpenStack platform into a single storage pool and provide a unified storage service API interface to the outside world, which appears as one large block of storage. If a user is only looking to perform some basic testing on the OpenStack platform, it is possible to forgo the installation of this service module. Typically, Cinder is used to meet the demands of Storage as a Service (SaaS), offering scalable storage solutions. Cinder provides enhanced extensibility, allowing users to request the necessary storage space without having to be concerned about the actual physical location where the data are stored.

(4) *Networking*: The Networking service module, known by the project name Neutron, is responsible for integrating all physical and virtual network resources within OpenStack. It provides a unified network configuration interface for constructing and configuring the networking environment of the OpenStack platform. Much like Nova, Neutron is also a core component of OpenStack.

(5) *Image Service*: The Image service module, known by the project name Glance, is a core component of OpenStack responsible for managing disk and server images and providing image services to VMs.

(6) *Object Storage*: The Object Storage service module, with the project name Swift, is another type of storage service within the OpenStack platform, similar to Cinder. Swift enables the storage and retrieval of various data types, including metadata and configuration data. It is also a suitable option for storing backup data, including data from the image service (Glance), which can be archived in Swift and retrieved through Glance when needed. Additionally, backups of block storage volumes managed by Cinder can also be stored in Swift.

(7) *Identity Service*: The Identity service module, with the project name Keystone, is a core component of OpenStack and serves as the system's identity service. Keystone handles authentication and authorization for users and services.

(8) *Monitor*: The Telemetry service module, known by the project name Ceilometer, is responsible for monitoring and metering the OpenStack cloud for billing, benchmarking, and statistical purposes.

6.2. The Compute Service Module Nova

Nova is a virtual server deployment and business computing module developed by the National Aeronautics and Space Administration (NASA), which is a virtualization management program capable of managing networking and storage. The primary implementation code for Nova is written in Python. Nova is the most essential module of OpenStack, with its functionality spanning nearly all areas of OpenStack. It acts as a computing controller, managing users' VM instances and performing operations such as starting and stopping VMs, as well as allocating CPU and RAM resources according to user requirements.

The Compute service module, Nova, leverages the Keystone module to provide authentication services, uses the Glance module to load and boot VM images, and the Neutron module to set up the network environment. Therefore, for Nova to operate correctly, its dependencies must be addressed before its installation. It is essential to ensure that the following modules are functioning properly prior to installing Nova: Keystone, Glance, and Neutron.

The Nova module is a substantial project. Due to its support for numerous interfaces and the extensive functionality it includes, Nova itself comprises many components. The important components of Nova and their workflow are illustrated in Figure 6.2.

(1) *Nova-API Component*: The Nova-API component serves as an interaction interface through which administrators can manage the internal infrastructure of OpenStack as well as provide services to users. Web-based management also operates through this interface by sending messages to the message queue to achieve resource scheduling functionality.

(2) *Nova-API-Metadata Component*: The Nova-API-metadata component is primarily responsible for handling requests for Metadata from VM instances. It is generally used under the multi-host mode with the Nova-network service installed.

(3) *Nova-Compute Component*: The Nova-compute component is in charge of managing the lifecycle of VM instances. It receives requests via the message queue and carries out operational tasks, making it a core service of Nova. The Nova-compute component typically runs on compute nodes, which are primarily responsible for data management, log management, configuration management, thread processing management, inter-process communication, TCP communication management, and more. It operates as a long-running daemon or background process on these compute nodes.

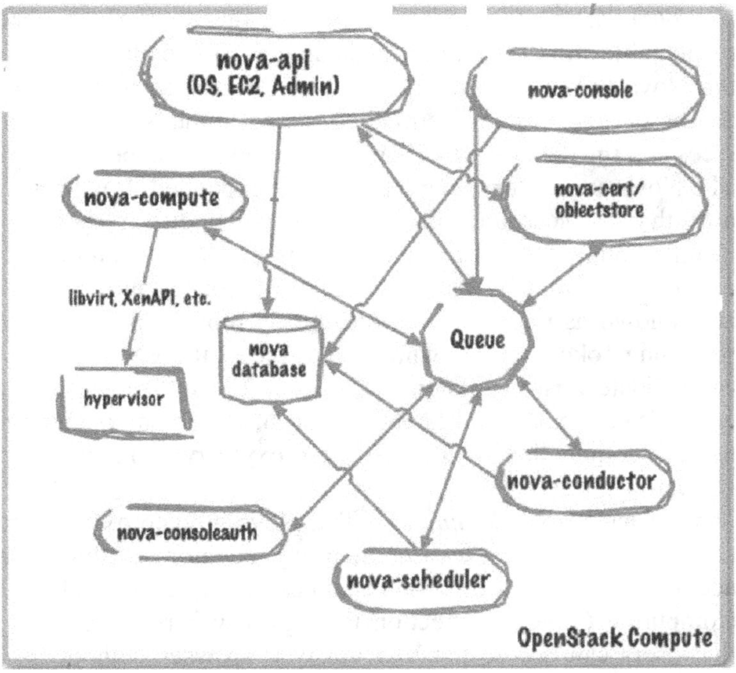

Fig. 6.2. The core components and working procedure of Nova.

(4) *Nova-Placement-API Component*: The Nova-placement-API component is primarily tasked with tracking and recording the inventory of resources available from Resource Providers, as well as the usage of those resources. It utilizes various resource classes to label resource types. Essentially, the Nova-placement-API component does nothing beyond information collection. It merely gathers data and provides corresponding APIs that allow users to view relevant statuses and usage details.

(5) *Nova-Scheduler Component*: The Nova-scheduler component acts as the brain behind the allocation of compute resources for new VM instances in OpenStack. It maps the calls coming from the Nova-API to the appropriate components within OpenStack, making decisions based on various factors such as CPU architecture, physical distance of available zones, memory capacity, current workload, and other customizable filters and weights.

(6) *Nova-Conductor Component*: The Nova-conductor component is responsible for mediating database access control, preventing the Nova-compute components from directly accessing the cloud database. It acts as an intermediary between the Nova-compute services and the database, facilitating communication between the two. The Nova-conductor component can be expanded. It is important to note that for security reasons, the Nova-conductor should not be installed on nodes that are also running the Nova-compute service.

(7) *Nova-Consoleauth Component*: The Nova-consoleauth component is also known as the Nova-consoleauth daemon. It is responsible for providing Token to users who wish to access the VM console. Token can validate a user's session and grant access to the console of a specific virtual machine. The Nova-consoleauth component must be used in conjunction with a console proxy service for it to function correctly.

(8) *Nova-Novncproxy Component*: The Nova-novncproxy component, also known as the Nova-novncproxy daemon, enables users to access their running VM instances using a Virtual Network Computing (VNC) connection through a web browser. It is noted that users can access the VM via web browser without installing VNC.

(9) *Nova-Spicehtml5proxy Component*: The Nova-spicehtml5proxy component, also known as the Nova-spicehtml5proxy daemon, provides users with the ability to access their VMs using the SPICE (Simple Protocol for Independent Computing Environments) protocol through an HTML5-compliant web browser. SPICE is an open remote computing solution that offers high-quality access to virtual desktop environments. It's designed for virtualized and cloud environments and can adapt to network conditions to deliver a user experience that closely matches that of interacting with a physical desktop. The Nova-spicehtml5proxy component serves a similar purpose to the Nova-novncproxy component in the OpenStack ecosystem, but it uses the SPICE protocol instead of VNC for the connection method. Both components provide web-based access to VM consoles through an HTML5-compliant web browser.

(10) *Nova-Xvpvncproxy Component*: The Nova-xvpvncproxy component, also known as the Nova-xvpvncproxy daemon, is another service within the OpenStack ecosystem that allows users to access their running VM instances using the VNC protocol. However,

unlike Nova-novncproxy, which uses HTML5 and WebSockets to enable access through a web browser, Nova-xvpvncproxy requires a Java applet to establish the VNC connection.

(11) *The Queue Component*: Nova consists of numerous subcomponents, each performing a specific subservice. These subservices need to coordinate and communicate with one another. To decouple the various subservices, Nova utilizes a Message Queue as an intermediary station for the exchange of information between subservices. Consequently, in the architectural diagram, there are no direct connections between subcomponents; instead, they are all linked through the Message Queue. The "queue" component is used for communication and message relaying between different servers. The functionality of the queue component is typically implemented using the RabbitMQ service, but other services can also be employed, such as the Advanced Message Queuing Protocol (AMQP) message queues realized through ZeroMQ.

(12) *SQL Database Components*: The SQL Database component is used for storing information, including records of the cloud platform's creation and operational status information. This encompasses details such as the types of VMs currently supported by the OpenStack platform, instances of VMs in operation, the status of network operations, and project performance. In theory, the OpenStack cloud platform supports all databases compatible with SQL Alchemy. However, during testing and development, databases such as SQLite3, MySQL, and PostgreSQL are commonly used.

6.3. Network Service Module Neutron

The network service module, Neutron, exists independently within the OpenStack architecture, much like Nova. It is used to assist other OpenStack modules in completing their tasks, such as compute services, image services, authentication services, and web services. Similar to other OpenStack components, Neutron is often deployed across multiple hosts.

The Neutron module within the OpenStack project manages all network ports, both virtual and physical. It integrates the entire network infrastructure, providing a unified API interface for the use of network resources. The capabilities of Neutron extend beyond mere management and provision of network interconnectivity; it also offers a variety of other useful and powerful features, such as firewalls, load balancing, and VPN services. For the network service platform to function correctly with Neutron, at a minimum, one external network and one internal network are required. The external network facilitates access from outside the OpenStack platform, while the internal network is used for communication between the various components within the OpenStack platform.

The external API interface of Neutron is provided by the neutron-server daemon, which is also responsible for managing the configuration of network function plugins.

When designing an OpenStack deployment, if the plan is to use a control host to centrally manage compute node components, Neutron can indeed be installed on this control node. However, it is not mandatory to deploy it on the control node because Neutron operates independently and can be installed on any host within the infrastructure. The specific installation and deployment of Neutron should be configured according to the particular requirements of the environment.

6.3.1. *Main components of neutron*

Neutron is similar to Nova, which is also composed of many components. The main components are introduced as follows:

(1) *Neutron Server* (*Neutron-Server and Neutron-*-Plugin*): The neutron-server is one of the core components of Neutron and is responsible for directly receiving external requests and then invoking the corresponding backend plugin for processing. The neutron-server primarily consists of the neutron-server daemon and various plugins referred to as neutron-*-plugin. It is important to note that both the neutron-server and all neutron-*-plugin services must run on a network node, which can be a dedicated network node or a control node. They provide the network module API interface and its extensions to external clients. Additionally, neutron-server is responsible for presenting the network model as well as information about IP addresses on each port.

(2) *Plugin Agent* (*Neutron-*-Agent*): The plugin agent, named neutron-*-agent, is responsible for handling packet processing on the virtual network. The term "plugin" typically refers to those residing within the neutron server and includes both the core plugin and service

plugins, whereas "agent" refers to components that run on individual nodes and are responsible for implementing network services. The operation of a plugin agent requires access to a message queue and the relevant dependent plugins to be pre-installed or available. However, it should be noted that some network technologies, such as OpenDaylight (ODL) or Open Virtual Network (OVN), do not require any plugin agents to be installed on the compute nodes.

(3) *DHCP Agent (Neutron-DHCP-Agent)*: The DHCP agent in OpenStack Neutron provides Dynamic Host Configuration Protocol (DHCP) services to tenant networks, which are networks created and managed by tenants within the OpenStack environment. The DHCP agent is responsible for dynamically allocating IP addresses to VMs within these networks. In addition to IP address allocation, the DHCP agent also facilitates metadata service requests. It's important to underline that the metadata request service spans across all plugin services and maintains DHCP configuration information. Like other Neutron agents, the DHCP agent requires access to the message queue.

(4) *L3 Agent (Neutron-l3-Agent)*: The L3 agent in OpenStack Neutron, which stands for Layer-3 Networking Extension, functions as an extension of the API and provides routing and Network Address Translation (NAT) services. This agent enables tenant networks to access external networks. The L3 extension comprises two types of resources. One is the router, which forwards data packets among various internal subnets and performs NAT via a designated internal gateway. Each subnet corresponds to a port on the router, and the IP address of this port serves as the gateway for the subnet. The other type is the floating IP, which represents an IP address from an external network, mapped to a port on the internal network. A floating IP can only be defined when the router:external attribute of the network is set to True.

(5) *Network Provide Services (SDN Server/Services)*: Network provider services offer additional network functionalities to tenant networks. These supplementary network services are facilitated by Software-Defined Networking (SDN) servers or services, which communicate with the neutron-server, neutron-plugin, and plugin-agents through communication interfaces, such as REST API endpoints.

It is important to note that for all of Neutron's agents, communication with the main network service is accomplished either using a Remote

Procedure Call (RPC) protocol, such as RabbitMQ or Qpid, or through standard API communications. Additionally, the integration of Neutron with various OpenStack modules can be achieved in many different ways and involves various dependencies:

(1) Neutron relies on Keystone to authenticate and authorize all API requests.
(2) Nova communicates with Neutron through standard APIs. For instance, during the creation of a VM instance, nova-compute interacts with the network service API to attach each virtual network interface on the virtual host to the corresponding network.
(3) The integration of the web interface, which is the dashboard service module known as Horizon, with Neutron is also achieved through standard APIs. This integration allows both administrators and project users to create and manage network services through a web-based interface.

6.3.2. *Neutron network*

Neutron is responsible for managing all the Virtual Networking Infrastructure (VNI) within the OpenStack environment, effectively serving as the network service component. Neutron abstracts the Physical Networking Infrastructure (PNI) components, such as networks, subnets, ports, and routers, into a virtualized form. VMs launched subsequently can then connect to this virtual network. With the foundational knowledge established, further introduction involves the deployment of OpenStack network services within the physical network infrastructure. A standard OpenStack network architecture typically divides the networking into four distinct and separate networks, as illustrated in Figure 6.3.

(1) *Management Network*: The Management Network in an OpenStack deployment is used for internal communication between OpenStack modules (services) and for accessing API endpoints. For security reasons, the management network should be isolated and restricted within the data center.
(2) *Guest Network*: Customer networks, also referred to as data networks, are utilized for data communication between VM instances deployed

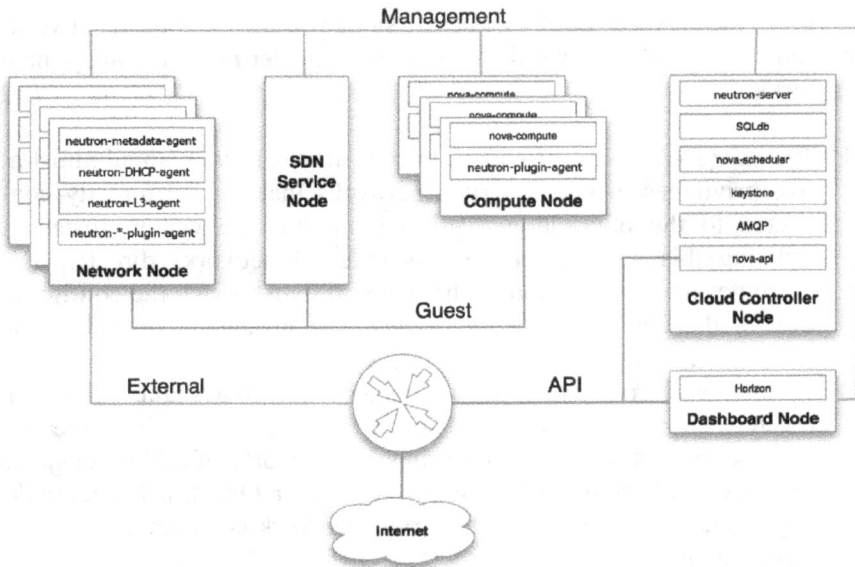

Fig. 6.3. Physical network structure of neutron.

on cloud platforms. The assignment of IP addresses within customer networks is contingent upon the plugin agents utilized by the OpenStack networking component Neutron, as well as the configuration of the virtual networks selected by the tenants. Customer networks are typically deployed within the secure domains of the customers.

(3) *External Network*: An external network, also known as a public network, is a network that can be accessed via the Internet or external connections. In certain deployments, the external network is used for allowing VM instances to access the Internet. The IP address of an external network is a publicly accessible address, and it is typically deployed in a public security zone.

(4) *API Network*: The API network ensures that all OpenStack APIs accessible to tenants are reachable, including those of the OpenStack networking module. The IP addresses in this network, akin to those in the external network, remain accessible to all connection requests originating from the public internet. Generally, the API access network creates a subnet within the external network, which typically resides in a public security zone.

From the perspective of user permissions within OpenStack, networking can be divided into two distinct types: provider networks and tenant networks.

(1) *Provider Network*: A provider network allows users with administrative privileges to create virtual networks that can be directly connected to the physical network within the data center. Instances utilizing this network can access external networks directly. The attributes and information of this network must match the configuration of the physical network because the instances connect to the external network through this network.

(2) *Tenant Network*: A tenant network is a network that regular users can create within their own tenant space. This network is invisible and isolated from other tenants, ensuring that networks of different tenants cannot directly communicate with one another. OpenStack's networking service, Neutron, supports various network isolation and overlay types, including Flat, Local, VLAN, GRE, and VXLAN.

6.4. Block Storage Service Module Cinder

Cinder is the block storage service within the OpenStack platform, designed specifically for end users, primarily to be used by VM instances managed by the Nova module. In essence, Cinder provides block storage services for virtualization environments. It also offers an API for self-service capabilities to end users, allowing them to extend their storage space on demand without the need to be concerned about the underlying data or hardware specifics.

Cinder, in the context of OpenStack, is typically employed for SaaS applications. Utilizing Cinder as the storage service comes with numerous benefits, such as the following: component-based architecture, which allows for rapid integration of new features; high availability and scalability, which help handle heavy loads by adding more instances; fault tolerance, which enables the prevention of cascading failures by isolated processing methods; recoverability, which allows for tracing and tackling errors.

Similar to other modules within OpenStack, Cinder needs to be installed on one or more nodes. While providing block storage services, Cinder also offers functionalities such as volume snapshots and volume types. The primary components of Cinder include the block storage API

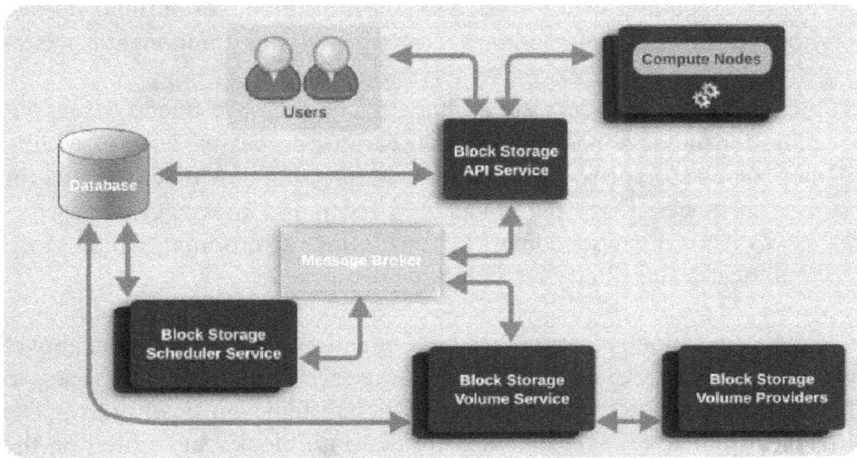

Fig. 6.4. Architecture framework of cinder.

service, block storage volume service, and block storage scheduler service. The architectural framework of the Cinder module is illustrated in Figure 6.4.

(1) *Block Storage API Service* (*Cinder-API*): The block storage API service, known by the project name cinder-API, is a component responsible for receiving requests and forwarding the contents of these requests to the cinder-volume component to perform the actual operations.

(2) *Block Storage Volume Service* (*Cinder-Volume*): The block storage volume service, whose project name is cinder-volume, interacts directly with storage services, known as block storage volume providers, to carry out operations such as the allocation and reclamation of storage blocks. Additionally, it is required to interact with the cinder-scheduler to execute tasks assigned by the scheduler. Finally, it must communicate with the message queue to process messages. The cinder-volume service responds to read operations and sends write requests to the block storage service to maintain state. Through its corresponding driver architecture, cinder-volume is capable of interfacing with a wide array of storage solutions.

(3) *Block Storage Scheduler Service* (*Cinder-Scheduler*): The project name for the block storage scheduler service is called

cinder-scheduler, which is used to select a suitable or optimal storage node to create storage volumes. Functionally, this component is somewhat similar to the nova-scheduler component.

(4) *Messaging Broker* (*Message Queue*): The message queue is responsible for message passing between various processes, such as adding new block storage to the platform or requesting new block storage for VM instances. Once the cinder-API receives a message, it then forwards it to cinder-scheduler for scheduling, determining which block to allocate, and finally returning the result.

The term "block storage" has been mentioned multiple times above; what exactly is block storage? Block storage provides users with access to storage devices based on data blocks, which differs significantly from file storage. In block storage, data are divided into blocks and stored on the device in indivisible units, similar to having an external hard drive. Files, on the other hand, need to be stored within these blocks. Users do not need to concern themselves with the physical location of the blocks; they can perform operations such as reading, writing, and formatting on the blocks. With block storage as the foundation, if a Logical Volume Manager (LVM) is available, users can easily expand their storage space.

6.5. Object Storage Service Module Swift

In OpenStack, Swift is also known as the Object Storage project. It serves as a cloud storage software that facilitates the retrieval and storage of multiple data through a simple API. Swift can enhance and optimize the persistence, availability, and concurrency of entire datasets. It is particularly well suited for storing unstructured data, relieving users of concerns regarding potential issues stemming from large volumes of data.

Similar to other modules in OpenStack, Swift conducts data read and write operations through a series of REST (Representational State Transfer, a web service architecture style where data are accessed and transferred using HTTP) calls and APIs (Application Programming Interfaces, which define a set of architectural principles for designing web services centered around system resources, including how clients written in different languages handle and transfer resource states). Swift consists of numerous components, and its architectural structure is illustrated in Figure 6.5.

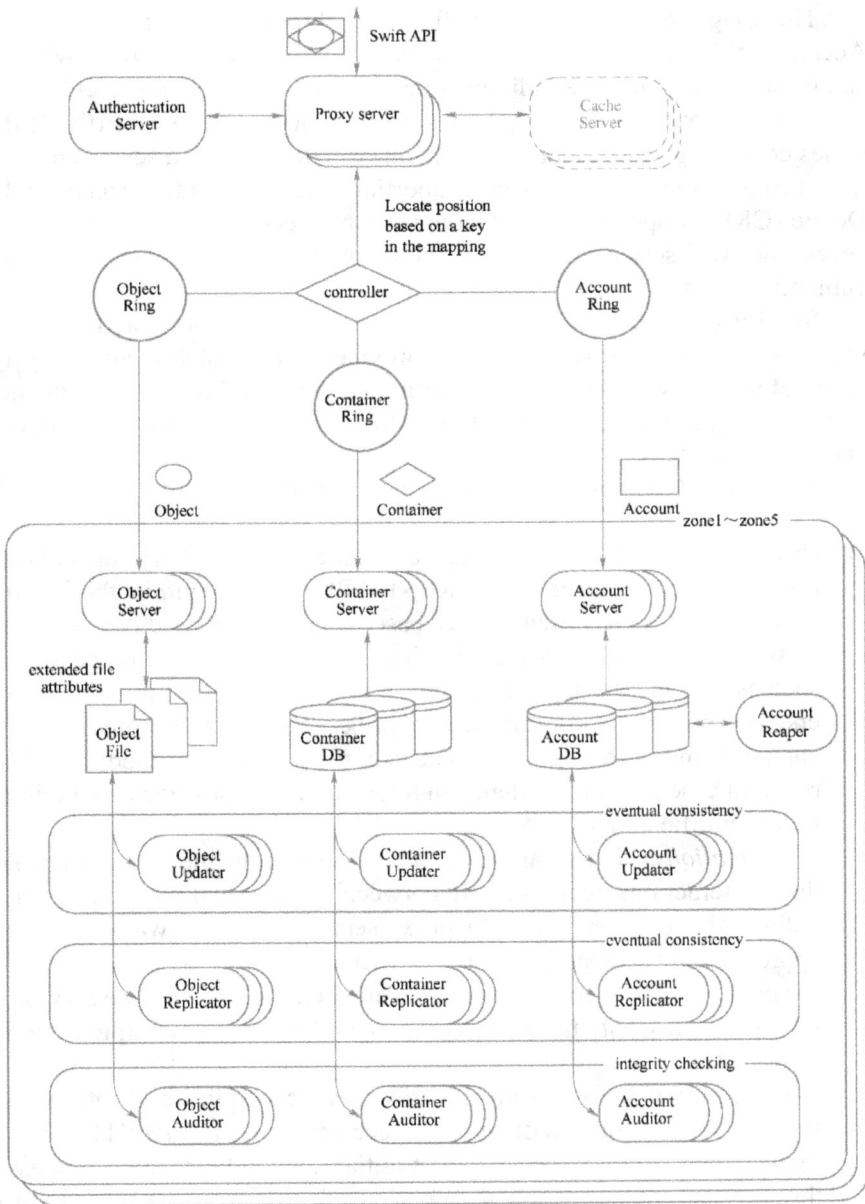

Fig. 6.5.   Architectural structure of swift.

The diagram shows that Swift has a three-tier logical structure: Account (akin to a tenant or user account for top-level isolation, which can be shared by multiple individual accounts), Container (representing a group of objects, similar to a folder or directory), and Object (the leaf nodes consisting of metadata and content). Swift provides a service interface through the proxy server, supporting Create, Read, Update, and Delete (CRUD) operations on accounts, containers, and objects. Prior to accessing Swift services, users must first obtain an access token from the authentication server.

Swift is designed with a fully symmetrical, resource-oriented distributed system architecture, where all components are scalable, allowing it to avoid system-wide disruptions due to single-point failures. It employs non-blocking I/O for communication, which enhances system throughput and responsiveness.

Now, let's delve into the main components of Swift.

(1) *Proxy Server*: The proxy server acts as an intermediary, providing proxy services to receive all REST API requests within the Swift framework. It routes each request based on its content and destination, forwarding it to appropriate handling modules for further processing, such as locating the appropriate entry point for user accounts, containers, or objects. Additionally, the proxy server is responsible for encoding and decoding object data according to Erasure Coding (EC) rules. In case of failures, it aids in returning error messages, including failure notifications.

    Attentions: If data arrive at or are sent from the object server, these interactions occur directly between the user and the data server, without the involvement of the proxy server for data forwarding. This approach enhances both transmission efficiency and security. Even if the proxy server encounters an issue and is unable to provide service, it will not disrupt the other components' ability to complete data transmission.

(2) *Authentication Server*: Authentication Server supplies authentication service. When the Swift system receives a user's RESTful API request via the proxy server, it first initiates an authentication process through the authentication server to verify the user's identity. Only after successful authentication does the proxy server proceed to handle the user's request and provide a response.

The authentication server verifies the user's identity information and generates an object access token (Token). Once authenticated, the server issues the token to the user. Subsequently, the user must include this token in every HTTP request they make. Swift, upon receiving the token and username, verifies its validity. If the token is valid, it proceeds with the requested operation; otherwise, it returns an authentication failure. Tokens are valid for a certain period; their validity is checked and cached until they expire.

(3) *Cache Server*: The cache server provides caching services, caching information about object service tokens, accounts, and containers but not the actual data of the objects. The cache server can be set up as a Memcached cluster, and swift employs consistent hashing to distribute cache addresses.

(4) *Ring*: Ring is used to maintain a mapping between the names of data stored on disk and their physical locations. A separate ring is assigned for each storage strategy, mapping accounts, containers, and objects. When other modules need to perform any operation on an Object, Container, or Account, they interact with the corresponding ring to determine the exact storage location within the cluster.

Ring employs zones, disks, partitions, and data replicas to facilitate mapping. In the context of ring, each partition is a replica, with three identical copies typically stored in the cluster by default. The locations of these partitions are recorded in a mapping table maintained by ring. In case of a failure, ring is responsible for determining which device should take over message handling and continue providing service.

Whenever possible, replicas within each partition are isolated to as many distinct zones, regions, services, or disks as feasible in order to maximize data resilience and fault tolerance.

Ring evaluates the weighted value of each device to determine its load capacity within the cluster, distributing data across various nodes. The weighted values can be used to achieve even distribution of partitions across disks, ensuring load balancing across the cluster. This feature is particularly useful, as it allows for the application of weighted values in scenarios where there is excessive data processing pressure or failures, such as during rebalancing in ring, where the number of partitions to be moved is controlled based on their weighted values.

(5) *Storage Policy*: Storage policy breaks away from the traditional approach where storage policies were determined by the designers and implementers, allowing users to set different replica counts, erasure coding parameters, storage media with different performance, locations, and backend storage devices on a per-container basis for data with varying needs. This fully embodies Swift's "Software-Defined Storage" philosophy. Each disk in the system is associated with one or more storage policies. Storage policies are realized through the use of multiple object rings, as each policy has its own independent ring. These policies may differ due to their association with specific hardware compositions, resulting in unique characteristics.

By default, users employ a three-replica configuration in Swift. However, a user can create a new storage policy, which can then be applied to a new container with a two-replica configuration. Additionally, if the user adds some SSDs to storage nodes, they can create a specific storage policy for certain containers, allowing data to be stored in these SSDs based on the defined policy.

Each container can be created with a specific storage policy at the time of its inception, and that configuration remains in effect throughout the container's lifetime. Once a container is created with a particular configuration, all data objects stored within it adhere to that policy for storage.

(6) *Object Server*: The object server offers object services, which is a straightforward data storage service that enables storing, retrieving, and deleting data locally on disk. Objects are stored in binary format in the file system, and metadata is saved in the file's extended attributes (xattrs). Naturally, extended attributes require support from the underlying file system chosen by the object server. Some file systems, such as ext3, do not natively support changing xattrs by default.

Each object has a unique path for storage, derived from the hash of the inherited object name and a timestamp. Storage operations are based on the latest stored version, ensuring that the most recent one is preserved. It's worth noting that deletion also creates a version of the file, albeit an empty one with a size of 0, ending in .ts (timestone). This ensures that the deleted file can be properly removed from the copy as well.

(7) *Container Server*: The container server is responsible for providing container services, which include managing container metadata, statistics, and maintaining a list of objects within it. The container server is unaware of the exact location of each object, only knowing which objects are stored within a specific container. Container information is stored in an SQLite database. The container server keeps track of the total number of objects and the overall storage usage of the objects it contains.

(8) *Account Server*: Account server is responsible for providing account services, which include delivering account metadata and statistics and maintaining a list of containers it contains. Each account's information is stored in an SQLite database. The functionality of account server is similar to container server, but it solely focuses on listing the available containers, not the data objects.

(9) *Updater*: The updater's function is to serialize tasks for delayed updates when objects are unable to be updated promptly due to high load. These tasks are queued in the local file system for asynchronous updating once the service recovers. Occasionally, container server or account server may not be able to update data in real time, especially during system failures or periods of high load. In such cases, if an update operation fails, it is stored in a local file system list, where the updater handles and retries the failed updates. For instance, suppose a container service is already loaded normally, and a new data object is added to it, but the service hasn't updated its object list at that moment. In this scenario, the container would appear as an empty one without any data objects. However, typically, once the proxy server sends a successful message back to the client, the data object's state should be marked as available and accessible. In this case, the update operation for the container would be queued for processing. The updater would then scan the queue and perform the necessary updates once the system returns to normal.

(10) *Replicator*: Replicator provides replication services, enabling the system to continue functioning normally in the event of failures, such as when network connectivity is lost or disk failures occur.

Replicator has two tasks: One is to perform redundant operations, where it backs up remote data and compares them with the local data to ensure that each copy is the latest version. Replicator can update object backups by synchronizing files to the other end. For account and container backups, it records missing data items and performs a complete database synchronization via HTTP or rsync.

The other task of replicator is to ensure that, if there's an object deletion operation, the marked deleted object is removed from the file system. In other words, once replicator detects a deletion marker from TomStone, it makes sure that the object is properly deleted.

(11) *Auditor*: The auditor is responsible for providing auditing services, verifying the integrity of data objects (Objects), containers (Containers), and accounts (Accounts). If bit-level errors are detected, the affected files are isolated, and a copy is made to replace the damaged local file. For other types of errors, log entries record the issue. If a list of objects expected to be found in a container server is missing, this is also logged to track the error.

6.6. Identity Module Keystone

The Identity module of OpenStack, known as Keystone, consists of various components to facilitate its services. Its primary purpose is to provide authentication and authorization functionality within the OpenStack cloud platform, ensuring a unified authentication process for all its components (projects). Keystone verifies the security of a user's identity and generates a globally unique access token (Token) through trusted internal services, which is then returned to the client. Subsequent access requests made by the client use this token for authentication. For example, when a client tries to access Nova, they present the token to the Nova-API. Nova receives the request and uses the token to make an authentication request to Keystone. Keystone verifies the token's authenticity by comparing it and checking its expiration date, ultimately responding with the validation result to Nova.

(1) *Various Versions of Keystone*: In terms of versions, Keystone does not have a V1, as its actual version predates OpenStack. The initial version of Keystone offered a limited set of APIs, which were implemented and provided by Rackspace, along with reference documentation. In the recent two versions of OpenStack, Keystone has been at

V2 and V3, and there are significant differences in their mechanisms.

- *Keystone V2*: In the V2 version of Keystone, a unique token is generated for each user using universally unique identifiers (UUIDs). However, this approach led to issues with online validation. With each user application request, Keystone verifies the client's UUID, placing a significant load on the server. During periods of high network traffic and when many users are simultaneously performing this verification, Keystone may struggle to respond in a timely manner, resulting in service disruptions or unavailability due to prolonged response times.

- *Keystone V3*: The V3 version of Keystone employs a more advanced authentication mechanism, called Public Key Infrastructure (PKI). Each token represents a key pair consisting of a public and private key for authentication and verification. With the advent of V3, the Keystone service gradually shifted to this version. Servers running Keystone are changed into a CA-like server, digitally signing all tokens. Since each token has its own data signature, it can be verified by any service, eliminating the need for OpenStack application components to query the Keystone database. In a similar manner to how a browser verifies the server's public key during an HTTPS connection, programs within the OpenStack project can check users' requests, ensuring secure communication.

    During the signature verification process, some information about the user's role is obtained, which is then used by OpenStack applications to decide whether to proceed with or reject the user's request. Any request will be denied if the signature is forged or has expired (the expiration time is also retrieved during the verification). Additionally, the second token in each user request is validated against a Certificate Authority (CA) revocation list. If a user's token is deleted by Keystone service or an administrator, it is automatically added to the CA's revocation list, and any application task in OpenStack carrying that token will be rejected.

    At present, OpenStack supports both V2 and V3 versions, utilizing similar processing logic. Before diving into the framework and general data flow of Keystone, let's first introduce its fundamental components and concepts.

(2) The basic components and certification process of Keystone:

- *Service*: Keystone is composed of a set of internal services (Services), which expose one or more endpoints (nodes or interface points) externally. For front-end services, Keystone internally combines multiple services to accomplish a function. For example, when handling an authentication request, Keystone verifies the user's and project's authentication information, utilizing the Identity service. After a successful verification, it calls the Token service to generate and return a Token to the user.

- *Identity*: The Identity service provides authentication for identity credentials and validation of user and group data. Typically, data access permissions are determined by the Identity service. If authorized, users can perform any operation on the data, such as CRUD.

- *User*: User (API user) is an individual API user. The user must belong to a specific domain because usernames may not be globally unique, but they are unique within their respective domain.

- *Group*: Group represents a container that can hold users, containing one or more users. Similar to users, groups must also belong to a specific domain, as group names are not globally unique, but they are unique within their domain.

- *Resource*: The Resource service provides data about projects and domains.

- *Project*: A Project is referred to as a Tenant. Projects in OpenStack represent the fundamental units with ownership relationships. All resources in OpenStack are owned by a specific Project. A Project must belong to a particular domain because project names are not globally unique, but they are unique within a domain. If a Project doesn't have a specified domain, it belongs to the default domain.

- *Domain*: A Domain is a high-level container for Project, User, and Group. There can only be one Domain per domain, and each Domain has its own namespace. Keystone typically provides a default domain named "Default". When using the V3 Keystone API, the uniqueness attribute of namespaces is as follows: *Domain Name* (globally unique); and *Role Name, User Name, Project Name, Group Name* (domain-scoped uniqueness). "Domains" could be used to manage OpenStack resources based on the container architecture, and users within a domain would still have access to resources within another domain, as long as they have the appropriate permissions.

- *Assignment*: The Assignment service provides data about Role and Role Assignments for the entities managing Identity and Resource Services.
- *Role*: Role represents the permission level which can be accessed by an authenticated user. The permissions of a role can be scoped at the domain or project level. Roles can be individually bound to a user, or they can be bound at the group level. Note that the role name is unique within a domain.
- *Role Assignment*: Role Assignment is used for assigning roles and is composed of a triple consisting of a Role, a Resource, and an Entity.
- *Token*: Generated and managed by the Token service, it is used for verifying and identifying user information.
- *Catalog*: The Catalog (directory service) provides endpoint registration, facilitating the discovery of endpoints.
- *Policy*: The Policy service offers a rule-based authentication engine along with a corresponding rule management interface.
- *Endpoint*: An Endpoint is a URL interface (REST API type) at the application frontend.

The following provides a brief overview of the authentication process in Keystone, assuming a scenario where a user, Alice, wants to create a VM instance. The steps in this process are illustrated in Figure 6.6:

(1) Alice sends the authentication credentials to the Keystone service. Upon a successful authentication, Keystone generates a temporary Token based on the provided credentials. This Token is then bundled with other relevant data, such as a version number, expiration time, and refresh period, to form an Endpoint that the user (Alice) can use to access the service. It's important to note that the Token is not permanent and has a limited lifespan.
(2) Alice gets the Token and sends a Keystone request to look at the project list she owns currently. Then, Keystone returns the list of projects related to the user.
(3) Alice provides her authentication credentials to Keystone and requests a specific project. Keystone verifies the credentials and generates a Token associated with the requested project (usually with a longer lifespan). It returns this project's current list of available services along with the created Token to the user. The service list contains the necessary Endpoint information for further access.

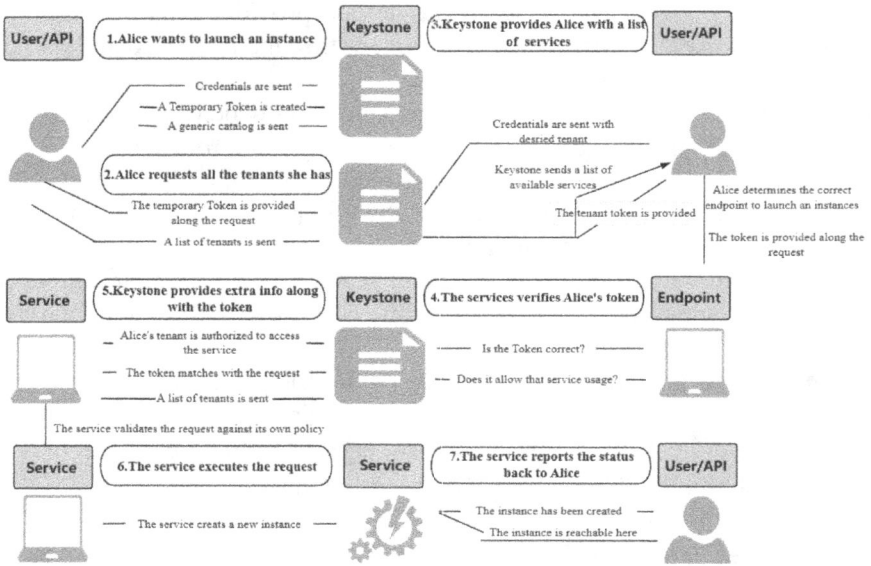

Fig. 6.6.   The authentication process of Keystone.

(4) After the Keystone service verifies the authentication, it returns the successful authentication details to Service A and also sends the user information linked to the Token (e.g., Alice's information if the Token is associated with her). If Alice's project has already been authorized in this service and the Token information matches the requested information (indicating it's Alice's Token), Service A receives the authentication confirmation from Keystone. Then, Service A validates Alice's request based on its own internal policy.

(5) After successful authentication by the Keystone service, it returns the authentication confirmation to Service A along with the user information associated with the token. If Alice's project has already been authorized in this service and the token information matches the request (e.g., the token belongs to Alice), Service A will validate the request initiated by Alice based on its own security policy configurations once it receives the authentication confirmation from Keystone.

(6) Upon completion of the execution, Service A sends a notification to Alice regarding the status of the requested operation, which may include but is not limited to "in progress", "successfully created", or

"failed". Following the operation, additional information about the newly created VM instance is also returned to the user, such as the accessible IP address. Users can then access the newly created VM instance using that IP address.

6.7. Image Module Glance

Glance primarily serves as an image service for VM creation. OpenStack is used to build a fundamental IaaS platform and offers VMs, and when a VM is being created, it requires an operating system to be installed. Glance is the component that provides various operating system images for the VMs being set up.

The Glance image service encompasses features such as discovery, registration, and retrieval of VM images. The images provided by the Glance module are usable by other OpenStack components such as Nova. With Glance, VM images can be stored in various storage options, like simple file storage or object storage systems (OpenStack's Swift project). Glance offers a REST API, allowing users to query metadata for VM images and retrieve them as needed.

Glance's design and implementation adhere to principles of component-based architecture (facilitating quick addition of new features), high availability (supporting heavy loads), fault tolerance, and open standards (providing a reference implementation for community-driven APIs). Like many other modules in OpenStack, Glance adopts a client-server (C/S) framework, providing a REST API to users for connecting and communicating with the service.

The Glance domain controller manages its internal service operations and hierarchizes them. Specific tasks are implemented within corresponding layers. The framework of Glance is depicted in Figure 6.7.

Glance utilizes a centralized database (Glance DB) to share information with all other components in the system, with a default implementation based on SQL.

(1) *A Client*: Any entity that connects to Glance, which could be an application in most cases.
(2) *REST API*: An API provided by Glance.
(3) *Glance*: Glance is also referred to as Glance Domain Controller. It acts as a middleware akin to a dispatcher, distributing Glance's internal service operations across six layers (Authentication, Notifier,

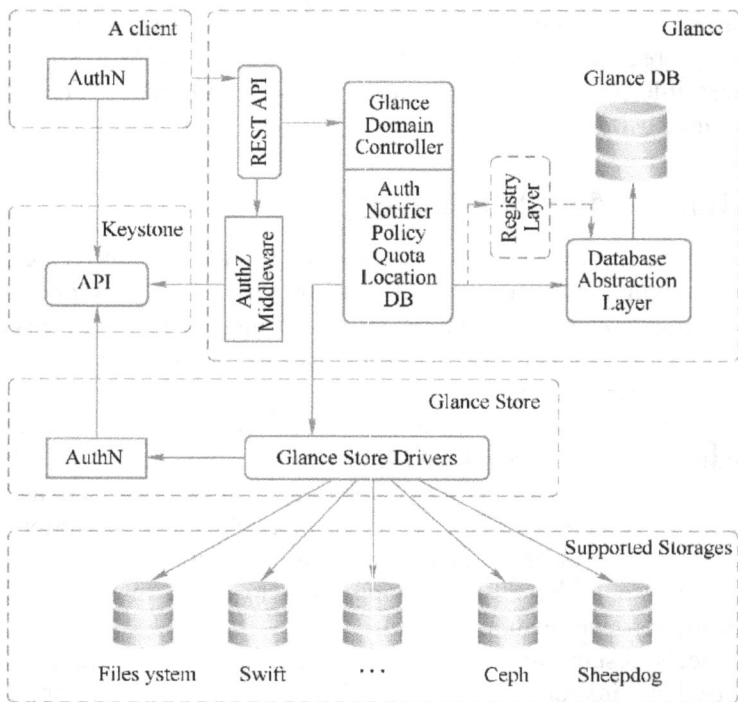

Fig. 6.7.   Framework of Glance.

Policy, Quota, Location, and DB connections). Each layer handles specific tasks. The Glance Domain Controller is responsible for implementing Glance's core functionalities, such as authentication, authorization, notification, rules, and database connections.

The functions of the six layers of the Glance Domain Controller are briefly explained as follows:

- Auth verifies whether the image or its properties can be modified. Such modifications can only be performed by administrators or the owners of the image; otherwise, the changes are saved.
- Notifier adds all notifications of image modifications and any exceptions or warnings that occur during usage to the queue.
- Policy defines access rules for the image operations, which are specified in the /etc/policy.json file, and monitors their execution.

- Quota is used to check if the user's upload exceeds the quota limit. If it doesn't exceed the limit, the operation to add the image is successful; otherwise, it fails and an error is reported. The administrator sets a quota for the size of all images that a user can upload.
- Location completes interactions with the Glance Store, such as uploading and downloading. Since the Glance Store can have multiple storage backends, it manages the locations where different images are stored. This involves verifying the correctness of a new image location when it is added, deleting the image from storage when its location is changed, and preventing duplicate image locations.
- DB (Database) implements interactions with the database API, converting images to the appropriate format for storage in the database. Additionally, it transforms the data received from the database into operational image objects.

(4) *DAL (Database Abstraction Layer)*: It is an API for application programming, acting as a communication interface between the database and Glance.

(5) *Registry Layer*: This layer may not be necessary, but it is added for security reasons. By isolating services, it ensures the safety of communication between the domain and the DAL.

(6) *Glance Store*: This component is responsible for facilitating communication between Glance and various backend data storage repositories (storage types), providing a unified interface for accessing the backend storage.

All operations on images are handled by the Glance Store, which is responsible for interacting with external storage interfaces. This external storage can be a local file system or a backend interface like the one provided by Swift. Regardless of the underlying storage method used by the backend, the interface exposed by the Glance Store remains consistent and unified to the outside.

6.8. Dashboard Service Module Horizon

In OpenStack, the Horizon dashboard service module provides users with a web-based Graphical User Interface (GUI) for managing OpenStack projects more conveniently. It enables users to monitor the operational status of the OpenStack platform in a more intuitive way, eliminating the need to remember numerous command-line interfaces. With Horizon, users can easily access information such as network topology, runtime status, CPU usage, memory usage, and disk usage. Figures 6.8 and 6.9 depict the Dashboard in an active/running state.

Figure 6.8 shows the login homepage of OpenStack, where the usage of cloud platform resources can be seen, such as the current number of cloud hosts, Virtual CPUs (VCPUs), and memory. Figure 6.9 is an example of OpenStack network topology, which illustrates the network deployment within the platform. As shown in Figure 6.9, there are two networks connected to the router my-route, net1 and net2, with gateways 10.10.10.1 and 20.20.20.1, respectively.

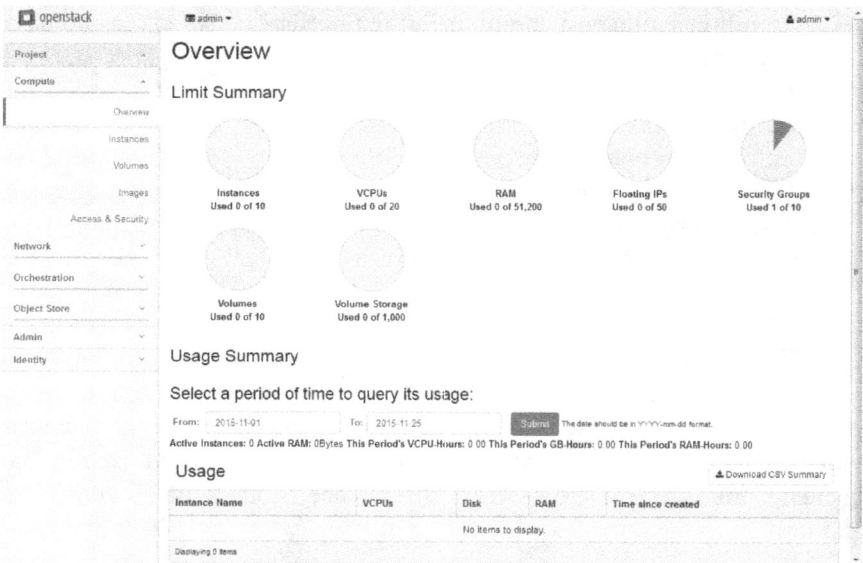

Fig. 6.8.   The page of dashboard after login.

Network Topology

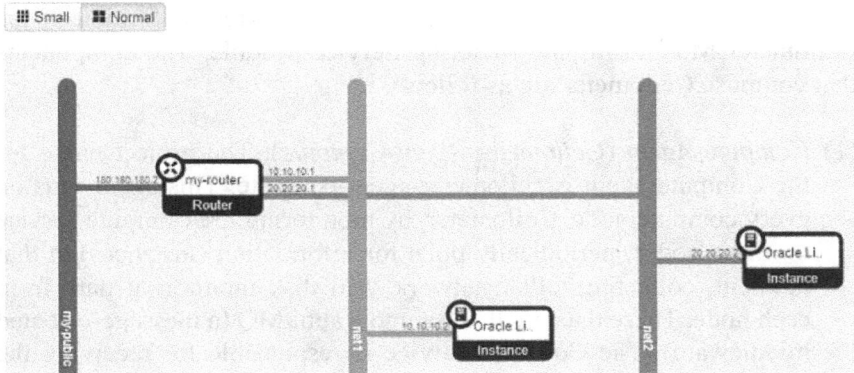

Fig. 6.9. Example of OpenStack network topology.

6.9. Monitoring and Metering Service Module Ceilometer

In OpenStack, Ceilometer functions like a funnel, collecting almost all events that occur within OpenStack and providing data support for billing, monitoring, and other services. Ceilometer can also gather information generated during operation and trigger messages. As the service runs longer, the amount of data collected by Ceilometer grows, and it issues warning messages to users or administrators when issues arise on the platform.

Ceilometer was initiated in 2012, initially designed to provide a foundational infrastructure for collecting all the information generated by the OpenStack platform, essentially serving as a framework for billing systems to gather data. However, as the project evolved, the scope expanded to include a broader range of information. Consequently, the OpenStack community broadened its goals, envisioning Ceilometer as the single-source infrastructure for data collection (including monitoring and billing data) within OpenStack. This data would be made available to projects, such as monitoring, billing, and dashboards, and could be organized into various formats as needed. This has led to the current definition of the Ceilometer Monitoring and Metering Service Module. The components that compose Ceilometer are as follows:

(1) *Compute Agent* (*Ceilometer-Agent-Compute*): The project name for the Compute agent is ceilometer-agent-compute. This agent runs on every compute node. Ceilometer, by monitoring the Compute service on these nodes, periodically polls for information (instances) at that moment, collecting CPU, network, and disk monitoring data from each node. These data are then sent to RabbitMQ (a message-oriented middleware). The Collector service is responsible for receiving the information, persistently storing it, and generating statistics.
(2) *Central Agent* (*Ceilometer-Agent-Central*): The project for the Central agent is ceilometer-agent-central. This agent runs on the central management server, where it periodically polls the APIs of various OpenStack components (such as Nova, Cinder, Glance, Neutron, and Swift) to gather resource usage statistics.
(3) *Notification Agent* (*Ceilometer-Agent-Notification*): The project for the Notification agent is ceilometer-agent-notification. This agent runs on either the central management or control server, where it retrieves notifications from OpenStack component message queues. The agent processes these notification messages and converts them into metering messages, which are then sent back to the messaging system. The metering messages are directly stored in the storage system.

Exercises

(1) Briefly summarize the main components of OpenStack?
(2) Briefly describe the functions of the Neutron networking service module.
(3) Briefly explain the functions of the Nova module.
(4) Briefly explain the differences between Swift and Cinder.

Chapter 7

Docker: A Versatile Container Technology

In Chapter 3 of this book, we introduced the basics of cloud computing, specifically virtualization technologies, and presented various virtualization solutions. Among these, container technology, represented by Docker, was classified as a direction within the broader category of virtualization. From a technological development perspective, container technology shares similarities with virtualization but is an evolution beyond it. Virtualization runs independent machines on physical hardware through an intermediary layer, while containers operate directly within the user space of the operating system kernel. This chapter systematically introduces Docker, a practical container technology, and explains its significant role in real-world applications.

7.1. Overview of Docker

Docker has three core concepts: Image, container, and repository. Containers are supported by images, while images are distributed from repositories and are built into rapidly deployable, diverse applications via pre-set commands.

7.1.1. *Installing Docker*

Docker officially supports installations on Windows, Linux, and macOS platforms. It also supports a visual client interface installation on Windows, significantly simplifying the installation process.

For a more practical work environment and to better understand Docker installation and usage in the command line state, this section explains the installation method on Linux CentOS. For other systems, please refer to the official documentation. The following are the specific steps for Docker installation:

(1) Linux systems usually come with Docker pre-installed, but the version may be too low, so you need to uninstall the old version first. The uninstallation command is as follows:

```
$ sudo yum remove docker \
docker-client \
docker-client-latest \
docker-common \
docker-latest \
docker-latest-logrotate \
docker-logrotate \
docker-selinux \
docker-engine-selinux \
docker-engine
```

(2) Execute the following commands to install the dependency packages, add a domestic source, and install docker-ce:

```
$sudo yum install -y yum-utils
$sudo yum-config-manager \
  --add-repo \
  https://mirrors.aliyun.com/docker-ce/linux/centos/
    docker-ce.repo

$sudo sed -i 's/download.docker.com/mirrors.aliyun.
  com\/docker-ce/g' /etc/yum. repos.d/docker-ce.repo

$ sudo yum install docker-ce docker-ce-cli containerd.
  io
```

(3) Start Docker:

```
$ sudosystemctl enable docker
$ sudosystemctl start docker
```

(4) Test whether Docker is installed correctly. If the following message appears, Docker has been successfully installed:

```
$ docker run --rm hello-world

Hello from Docker!
This message shows that your installation appears to
  be working correctly.

To generate this message, Docker took the following
  steps:
1. The Docker client contacted the Docker daemon.
2. The Docker daemon pulled the "hello-world" image
   from the Docker Hub.
   (amd64)
3. The Docker daemon created a new container from that
   image which runs the executable that produces the
   output you are currently reading.
4. The Docker daemon streamed that output to the
   Docker client, which sent it  to your terminal.

To try something more ambitious, you can run an
  Ubuntu container with:
 $ docker run -it ubuntu bash

Share images, automate workflows, and more with a
  free Docker ID:
 https://hub.docker.com/

For more examples and ideas, visit:
 https://docs.docker.com/get-started/
```

At this point, Docker has been installed and started successfully, running the hello-world container provided by Docker's public repository.

- Docker services are divided into client and server sides.
- When running an image directly, it is first looked up from the local repository or, if not available, from the Docker Hub public repository.
- This container run is for display-only; for interactive access to the container, use docker run -it ubuntu bash.

7.1.2. *Running the first container*

After running the hello-world container, you will gain a basic understanding of Docker's operation. To run an interactive container, use the following command:

```
docker run -i-t centos /bin/bash
```

The output of this command is quite rich, so let's break it down line by line.

First, the user tells Docker to execute the docker run command with the -i and -t options. The -i option enables standard input inside the container, and the -t option allocates a pseudo-TTY terminal to the container. This way, the new container can provide an interactive shell. If you want to create a container that you can interact with from the command line, these two options are necessary.

Next, Docker is told which image to use for creating the container—in this case, the CentOS image, provided by Docker and stored on Docker Hub. You can build your own images based on the CentOS base image (or similar Linux images). So far, a container has been started based on this image without any modifications to the container.

Now, what does the Docker daemon do with this command? The daemon uses the image to create a new container. The container has its own network, IP address, and a bridge network interface to communicate with the host machine, as shown in the following code:

```
[root@4b81bc62ac18 /]#
```

Then, Docker is told which command to run inside the new container. In this case, we are running the /bin/bash command, which starts a Bash shell in the container. This opens a pseudo-TTY terminal connected to the inside of the container and logs in as the root user. The container ID is 4b81bc62ac18. Inside the container, you can operate as if it were a regular Linux system.

You can perform any tasks you want inside the container. When you're done, typing exit returns you to the host machine's command prompt.

The container will stop running, but it will continue to exist. You can view the list of containers on the system with the following command:

```
[root@iZbp11i8w8iqclmzdvhnocZ ~]# docker ps -a
CONTAINER ID IMAGE COMMAND CREATED STATUS PORTS NAMES
4b81bc62ac18 centos "/bin/bash" 16 minutes ago Exited
    (127) 3 minutes ago upbeat_robinson
```

By default, the docker ps command only shows running containers. If you add the -a option, it will show all containers, both running and stopped.

From this output, we can see useful information about the container: its ID, the image used to create it, the last executed command, the creation time, and its exit status.

Docker automatically generates a name for each container, such as upbeat_robinson. If you want to assign a specific name, you can use the --name option. In many Docker commands, you can use the container name instead of the container ID to manage the container more easily.

Now, the upbeat_robinson container has stopped. You can restart it using the following commands:

```
docker start upbeat_robinson
docker start 4b81bc62ac18
```

7.1.3. *Basic Docker commands*

In the previous section, we detailed the process of starting a Docker container. Here, we summarize some basic Docker commands.

(1) The command to create a container is as follows:

```
Usage: docker run [OPTIONS] IMAGE [COMMAND] [ARG...]
-d, --detach=false        Run the container in the
                          background (default is
                          false)
 -i, --interactive=false  Open STDIN for interactive
                          input
```

-t, --tty=false	Allocate a TTY for terminal login (default is false)
-u, --user=""	Specify the user for the container
-a, --attach=[]	Standard input and output streams and error messages (must be a container started with non-docker run -d)
-w, --workdir=""	Set the working directory in the container.
-c, --cpu-shares=0	Set container CPU weights, used in CPU sharing scenarios
-e, --env=[]	Specify the environment variable, which can be used in the container
-m, --memory=""	Specify the upper memory limit of the container
-P, --publish-all=false	Publish all exposed ports
-p, --publish=[]	Publish specific ports
-h, --hostname=""	Specify the hostname of the container
-v, --volume=[]	Mount a storage volume to the container, to a directory in the container
--volumes-from=[]	Mounting a volume on another container to a directory on the container
--cap-add=[]	Add Permissions
--cap-drop=[]	Delete Permissions
--cidfile=""	After running the container, write the container PID value in the specified file
--cpuset=""	Set which CPUs can be used by the container, this parameter can be used to

	give the container exclusive access to CPUs.
`--device=[]`	Add a host device to a container which is equivalent to a device pass-through
`--env-file=[]`	Specify the environment variable file, the file format is one environment variable per line
`---expose=[]`	Specify the port to which the container is exposed, i.e., modify the exposed port of the image
`--link=[]`	Specify inter-container associations that use the IP, env, etc. of other containers
`--name=""`	Specify the name of the container, which can be used later for container management
`--net="bridge"`	Container network settings
`--restart="no"`	Specify the restart policy after a container stops
`--rm=false`	Specifies that containers are automatically deleted when they are stopped (containers started with docker run -d are not supported containers started with docker run -d)

(2) To view Docker containers, use the following commands:

```
docker ps     # View running containers
dockerps -a   # View all containers (running and
              stopped)
```

(3) Commands to start, stop, restart, and delete Docker containers are as follows:

```
#start docker
docker start <ContainerId(or name)>
#stop a container
docker stop <ContainerId(or name)>
#restart a container
docker restart <ContainerId(or name)>
#remove a container
docker rm <ContainerId(or name)>
#remove all container
docker rm $(docker ps -a -q)
```

For more commands, refer to Docker's official documentation: https://docs.docker.com/reference/.

7.2. Docker Images and Repositories

After starting your first Docker container, you should have a basic understanding of how containers are used. Next, we explain the principles behind running containers.

7.2.1. *What is a Docker image?*

To understand Docker more deeply, it is essential to grasp the concept of image layers. Image layers rely on a series of underlying technologies such as file systems, Copy-On-Write (COW), and Union Mounts.

Docker images are constructed by stacking layers of files. Similar to virtual machine images, Docker images employ a layered structure, allowing users' write operations to occur only in a dedicated read/write layer. When you build a container, the Docker daemon creates an image stack by pulling images from a repository and layering them. The images at the base are called base images, while any images above are called parent images. Once all public images are stacked, the daemon adds a read/write layer for user operations. The user interacts only with this read/write layer, while all the other layers are read-only.

A Docker image is a stack of filesystems. It is similar to an image in a virtual machine, but different in that it is layered so that user writes are only directed to a separate read/write layer generated for the user. Docker refers to such a file system as an image. When the user wants to build a container, the daemon generates a stack of mirrors and pulls them from the repository based on the user-specified mirrors and pushes them into the stack hierarchically, with the bottom mirror being called the parent mirror and the bottom mirror being called the base mirror. Finally, when all public mirrors are in the stack, the daemon generates a read/write layer in the stack for the user to operate. In this mirror stack, the user only interacts with the read/write layer, and the mirrors below it are read-only. Figure 7.1 shows the above structure.

When Docker first starts a container, the initial read/write layer is empty. When changes are made to the file system, those changes are applied to this layer. For example, if you want to change a file, that file is first copied from the read-only layer below that read/write layer. The read-only version of that file does not change but has been hidden by the copy of that file in the read/write layer, a mechanism known as COW. Each read-only mirror layer is read-only and never changes later. This read/write layer, together with the mirror layer below it and some configuration data, makes up a container.

7.2.2. Publishing and fetching Docker images

An important part of building images is how to publish and get them. This can be done by pushing images to Docker Hub or to a user's own private repository. Docker Hub is a cloud-based repository dedicated to building applications and maintaining containers. It provides a centralized repository for container image retrieval, release and change management, users and teams, and automation of the development process. To use the public

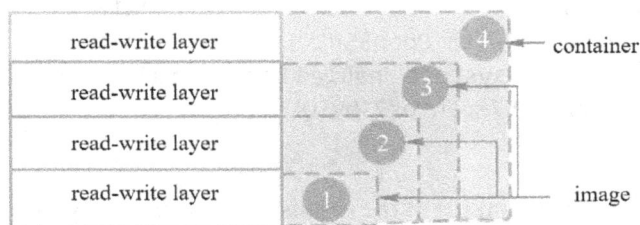

Fig. 7.1.   Container structure.

repository, first register for a Docker Hub account at https://hub.docker.com/ and then use the command to link this account to the Docker Hub public repository.

Execute the docker login command, enter the username and password, log in to Docker Hub from the command line interface, and log out using the docker logout command.

```
# docker login
Login with your Docker ID to push and pull images
    from Docker Hub. If you don't have a Docker ID,
    head over to https://hub.docker.com to create one.
Username:xxxxxxx
Password:
WARNING! Your password will be stored unencrypted in
    /root/.docker/config.json.
Configure a credential helper to remove this warning.
    See
https://docs.docker.com/engine/reference/
    commandline/login/#credentials-store

Login Succeeded
```

Users can use the docker search command to find images in the official repositories and download them locally using the docker pull command. For example, if you use centos as a keyword, you can see the following as an example:

```
docker search centos
NAME     DESCRIPTION    STARS   OFFICIAL    AUTOMATED
centos The official build of CentOS. 6695      [OK]
ansible/centos7-ansible Ansible on Centos7 134  [OK]
consol/centos-xfce-vnc Centos container...129     [OK]
jdeathe/centos-ssh OpenSSH... 119             [OK]
centos/systemdsystemd enabled base container. 101  [OK]
centos/mysql-57-centos7 MySQL 5.7 SQL database server
    91
```

This shows a partial list of the results of the search, as you can see that mirrors containing this keyword are searched and sorted by default by stars, i.e., popularity. Each line in the list contains basic information about the mirror, including the mirror name, description, popularity, whether it is officially created or not, and whether it is auto built or not.

In terms of official builds, one type of image is centos, which is created, verified, and supported by Docker, and has a unique image name. Another type of image is centos/systemd, which is created and maintained by Docker Hub users and is usually qualified by a "/".

When searching for a specific mirror that is needed, the user can specify parameters with --filter, where stars qualifies the popularity of the mirror, while automated and official are used to determine if the mirror is automatically built and officially maintained, respectively.

```
--filter is-automated=true --filter stars=3 --filter
  is-official=false
```

In addition, the search command has the following parameters available:

(1) *–format*: Formulate the fields to be displayed in the search list, as shown in the following example:

```
docker search --format "{{.Name}}:  {{.StarCount}}"
  centos
```

(2) *–limit*: Display the maximum number of bars.
(3) *–no-trunc*: Output the description field in its entirety.

Once you know how to find mirrors on Docker Hub, the next aspect to consider is how to pull the mirrors you need to build containers locally. When the user runs a container directly from the command line, Docker first checks the local registry for the required image. If the image is not found locally, it pulls the image from a remote registry (public or private). While this approach is convenient, it starts a new container every time the docker run command is executed, resulting in low efficiency. If you only want to pull the image locally, you can use the docker pull command, which has the following format:

```
docker pull [OPTIONS] NAME[:TAG|@DIGEST]
```

The name of a Docker image consists of "unique image name: version number". In the previous section, when building a container based on a centos image, you only entered centos as the unique image name, meaning that the pulled image is the latest version by default, which is actually the same as the image pulled by docker pull centos:latest. When you need a specific version of an image, the command should look like the following:

```
docker pull centos:7.6
```

In terms of optional parameters, due to fewer opportunities to use them, we will not introduce them here; interested readers can refer to the official documentation.

7.2.3. *Image operations*

After pulling the image locally, you can display the list of local mirrors with the docker image ls command, as shown in the following code:

```
# docker image ls
REPOSITORY   TAG     IMAGE ID   CREATED      SIZE
httpd alpine3.14 feb558a43c19 7 days ago     54.8MB
hello-world latest d1165f221234 5 months ago 13.3kB
centos latest 300e315adb2f 8 months ago     209MB
```

The list also provides the following information:

- *REPOSITORY*: the name of the mirror in the repository.
- *TAG*: image version/tag.
- *IMAGE ID*: the image ID.
- *CREATED*: the date the image was created (not the date it was fetched).
- *SIZE*: the size of the mirror.

During the image pulling process, multiple mirrors are downloaded in layers, and only the top layer of the final composition is shown in the mirror list viewed above. In addition to this, a special representation of mirrors may be seen in the mirror list, as shown in the following:

```
<none><none> 44a113458bad 7 days ago 2GB
```

There are two reasons for such mirroring. One is that the image has an original name and label, but the maintainer pushes different changes using the same label, then the name and label are transferred to the new image, and the original locally saved image becomes a virtual hanging image. Another situation is that when an image needs to use the existing top-level image as the middle layer, if you use docker image ls -a to view all the images, you will find more entries similar to the form of a virtual suspension of the image. This is because Docker uses a layered storage mechanism: each image layer is stacked on top of the base image, and only the differences between layers are stored. This approach minimizes redundant data and significantly reduces disk space usage.

Images that are no longer needed can be removed with the docker rmi or docker image rm commands, as shown in the following code:

```
# docker rmi nginx
Untagged: nginx:latest
Untagged: nginx@sha256:8f335768880da6baf72b70c701002
    b45f4932acae8d574dedfdda-
f967fc3ac90
Deleted: sha256:08b152afcfae220e9709f00767054b824361
    c742ea03a9fe936271ba520a0a4b
Deleted: sha256:97386f823dd75e356afac10af0def601f2cd
    86908e3f163fb59780a057198e1b
Deleted: sha256:316cd969204ae854302bc55c610698829c9f
    23fa6fcd4e0f69afa6f29fedfd68
Deleted: sha256:dcec23d16cb7cdbd725dc0024f38b39fd326
    066fc59784df92b40fc05ba3728f
Deleted: sha256:1e294000374b3a304c2bfcfe51460aa59923
    7149ed42e3423ac2c3f155f9b4a5
Deleted: sha256:c0d318592b21711dc370e180acd66ad5d42f
    173d5b58ed315d08b9b09babb84a
Deleted: sha256:814bff7343242acfd20a2c841e041dd57c50
    f0cf844d4abd2329f78b992197f4
```

When deleting an image, two types of output are commonly seen: *Untagged* and *Deleted*. This occurs because the unique identifier of an image is based on its ID and digest, and a single image can have multiple tags. When you delete an image, Docker first removes all the associated tags. If there are still other tags pointing to the image after the removal, the actual deletion of the image does not occur. Therefore, not all deletions will remove the image file itself, and whether or not the image is fully deleted depends on the presence of other tags.

Docker images are stored in layers; therefore, during deletion, Docker checks layer by layer from the top down. This multi-layer structure makes image reuse very efficient, as other images might rely on specific layers of the current image. In such cases, the dependent layer will not be deleted. Only when no other images depend on a given layer will Docker delete it. Besides image-to-image dependencies, containers can also depend on images. If there are containers that were created using a particular image (even if the containers are not running), the image cannot be deleted.

7.2.4. *Building a private repository*

Although Docker Hub offers a rich public repository, there are some practical challenges when using it. First, network speeds can be slow, and second, the public repository is open to everyone, which makes it unsuitable for storing images that require confidentiality. Therefore, learning how to build and use a private repository is essential.

Docker provides a tool called *Docker Registry* that allows users to build their own private repository. The way it operates is similar to running a Docker image. Here's the command to set it up:

```
docker run -d -p 5000:5000 --name registry registry:2
```

This command creates a local repository in the form of a Docker container named registry, binding the repository's port 5000 to the host machine's port 5000 for access. To test the usability of this private repository, we'll push a CentOS image pulled from the public repository to our private one.

Before pushing the image, it's necessary to assign it a unique tag to distinguish it from the officially maintained images. The command is as follows:

```
docker image tag centoslocalhost:5000/mycentos
```

Now, you can attempt to push the CentOS image to the private repository using the following command:

```
# docker push localhost:5000/mycentos
Using default tag: latest
The  push  refers  to  repository  [localhost:5000/
   mycentos]
2653d992f4ef: Pushed
latest: digest: sha256:dbbacecc49b088458781c16f3775f
   2a2ec7521079034a7ba499c8-b0bb7f86875 size: 529
```

At this point, Docker has successfully pushed the image to the private repository. In addition to building a private repository yourself, Docker also offers paid versions of private repositories. For more information, refer to Docker's official documentation.

7.3. Dockerfile—Customizing Images

In the previous two sections, we introduced Docker's three core concepts: containers, images, and repositories. In real-world applications, generic images provided by various companies often make software deployment much easier, but they may not be well suited for special needs due to their general-purpose nature. On the other hand, some research institutions may need to create new Docker images to promote their innovations. This is where Dockerfile comes in, allowing users to create highly customized images with more features.

7.3.1. *Dockerfile introduction*

As mentioned earlier, images are constructed in layers, with each layer gradually built upon the previous one. When deploying a container, commands are added to give the container new functionalities. The background daemon receives the command, adds an empty read/write layer, and executes the command, after which the container is started.

To simplify the image customization process, Docker provides a way to define the build process using a text file called Dockerfile. This file contains a series of commands used to build the image layer by layer.

A typical Dockerfile consists of four parts: base image information, maintainer information, image operation commands, and commands executed when the container starts. The following gives an example:

```
FROM centos
MAINTAINER mymy@email.com
RUN yum -yinstallnginx
RUN echo '<h1>Hello, Docker!</h1>' > /usr/share/
    nginx/html/index.html
CMD nginx
```

This Dockerfile is quite simple. It uses CentOS as the base image, installs the nginx web server, and changes the default page to display "Hello Docker!". Afterward, the image can be built using the docker build command.

7.3.2. *Explanation of Dockerfile commands*

In the previous example, four Dockerfile commands were used: FROM, MAINTAINER, RUN, and CMD, which correspond to specifying the base image, maintainer information, image operation commands, and startup execution commands, respectively. In the following, we discuss these image-building commands from these four aspects.

To create a custom image, in most cases, it is necessary to base it on an existing image, which aligns with Docker's multi-layer build philosophy. The FROM command specifies the base image, so it is a required command in any Dockerfile and must be the first command. There are numerous official and unofficial images on Docker Hub, some of which can be directly used for services, such as nginx and mysql. If none of these base service images meet your requirements, you can also choose images based on development languages and auxiliary tool libraries, such as JDK and Python, for more flexible builds. Lastly, if you need to perform pioneering or cutting-edge development, you can base your image on the more fundamental operating system images provided officially. Operating system images provide a broader space for building custom images. Additionally, Docker has a special image called scratch. This image is virtual and only exists to comply with Dockerfile build syntax. Using this image as a base means no image is used as the foundation, and the following commands will build the first layer of the image.

The MAINTAINER command specifies the maintainer's information.

Once the base image is selected, the RUN command is used to further modify the base image. Every time a RUN command is executed, an additional layer is added to the image. As one of the most commonly used commands for customizing images, the RUN command has two usage formats:

(1) *shell format*: RUN <command>, which is equivalent to the shell command format.
(2) *exec format*: RUN ["executable", "param1", ...], which offers more flexibility than the shell format.

It's important to note that you should try to combine multiple RUN commands into a single one and add the necessary files and environment. As mentioned earlier, every RUN command submits a new layer to the image, and running many separate commands will add a large number of unnecessary intermediate layers. This increases both the build time and the image size, and it also makes errors more likely. At the end of the build file, the CMD command specifies what command to execute when the container starts. The CMD command supports the following three formats:

(1) CMD ["executable", "param1", "param2"] (this format is recommended).
(2) CMD <command>.
(3) CMD ["param1", "param2"].

When the container starts, if the user specifies a command to run, it will override the predefined command. In a Dockerfile, only one CMD command will be executed, meaning if multiple commands are defined, only the last one will take effect.

7.3.3. *Building custom images*

In addition to the basic commands mentioned earlier, Docker provides more advanced commands to build diverse and customized images:

(1) *EXPOSE*: The format is EXPOSE <port> [<port>...]. This command tells the Docker daemon which port(s) the container will expose for

external systems to use. When starting the container, you need to use the -P option for Docker to automatically map a host port to the specified container port. It's important to note that the EXPOSE command only declares the port; it doesn't actually open it. If you don't use the -p or -P option to map the port, it won't be accessible externally.

(2) *ENV*: The format is ENV <key><value>. This command tells the Docker daemon which port(s) the container will expose for external systems to use.

(3) *ADD*: The format is ADD <src><dest>. This command copies files from a specified source to the container's destination directory. The source (<src>) can be a relative path in the Dockerfile's directory, a URL, or a tar file (which will be automatically extracted).

(4) *COPY*: The format is COPY <src><dest>. This command copies files from the local machine<src> (<src> refers to the relative path to the directory where the Dockerfile is located) to the container<dest>. When using a local directory as the source, it is recommended to use the COPY command.

(5) *VOLUME*: The format is VOLUME ["/data"]. This command creates a mount point that allows directories to be shared between the host machine or other containers and the container. It's often used for persistent data storage, like databases.

(6) *WORKDIR*: The format is WORKDIR /path/to/workdir. This command sets the working directory for subsequent RUN, CMD, and ENTRYPOINT commands.

(7) *ENTRYPOINT*: This command configures the commands that will always run when the container starts and cannot be overridden by docker run parameters. The format is the same as that of CMD.

These commands cover most of the essentials for building custom Docker images. After writing the Dockerfile as needed, you can use the docker build command to build the image. The docker build command requires specifying the context directory. During the build process, Docker will package the context and upload it to the Docker engine. Commands such as RUN that operate on the base image don't need to interact with the host environment. However, commands like COPY and ADD do interact with the host environment. Therefore, it's recommended to keep the context directory as minimal as possible to avoid unnecessary file transfers during the image build.

In addition to manually writing Dockerfiles, you can also build images using other methods, such as reading Dockerfiles from a network source or directly unpacking a given tar file, which greatly improves the flexibility of container migration and deployment.

7.4. Kubernetes—Container Orchestration Technology

Docker is one of the most widely used container technologies. While it provides an excellent solution for packaging and running services as containers, as applications grow more complex, managing and maintaining an increasing number of containers becomes challenging. Furthermore, as cloud computing evolves, managing vast numbers of containers in the cloud introduces significant operational complexity.

7.4.1. *Introduction to Kubernetes*

Kubernetes, also known as K8s, was originally designed by engineers at Google. It is an open-source platform that automates the operation of Linux containers. Kubernetes can help users automate many of the manual tasks involved in deploying and scaling containerized applications.

With Kubernetes, you can group and manage clusters of hosts running Linux containers. These clusters can be deployed across public, private, or hybrid clouds:

(1) *Service Discovery and Load Balancing*: Kubernetes can expose containers using an IP address and distribute traffic if the load is high.
(2) *Storage Orchestration*: It allows automatic mounting of storage systems, whether local or cloud-based.
(3) Kubernetes efficiently manages the deployment, scaling, and updates of applications.

7.4.2. *Deploying Kubernetes*

To learn how to use Kubernetes, the user must first deploy it. You need at least two servers or virtual machines with dual-core CPUs running at

4 GHz or higher to set up a multi-node Kubernetes cluster. First, install Docker on each machine using the following commands:

```
# Tools needed to install docker
yum install -y yum-utils device-mapper-persistent-
  data lvm2
# Configure AliCloud docker sources
yum-config-manager --add-repo http://mirrors.aliyun.
  com/docker-ce/linux/
centos/docker-ce.repo
# Specify to install the specified version of
  docker-ce
yum install -y docker-ce-18.09.9-3.el7
# start docker
systemctl enable docker &&systemctl start docker
```

Next, disable unnecessary services and configure system parameters as follows:

```
# disable the firewall
systemctl disable firewalld
systemctl stop firewalld
# disable selinux
# disable temporarily selinux
setenforce 0
# permanently disable modifying /etc/sysconfig/
  selinux file settings
sed -i 's/SELINUX=permissive/SELINUX=disabled/' /
  etc/sysconfig/selinux
sed -i "s/SELINUX=enforcing/SELINUX=disabled/g" /
  etc/selinux/config
# disable the swap partition
swapoff -a
# To disable it permanently, open /etc/fstab and
  comment out the specified swap line
sed -i 's/.*swap.*/#&/' /etc/fstab
# modify kernel parameters
cat <<EOF > /etc/sysctl.d/k8s.conf
```

```
net.bridge.bridge-nf-call-ip6tables = 1
net.bridge.bridge-nf-call-iptables = 1
EOF
sysctl --system
```

Then, install the necessary Kubernetes components: kubeadm, kubectl, and kubelet:

```
# Due to network environment issues, the configuration
  of the Kubernetes AliCloud source is performed here
cat <<EOF > /etc/yum.repos.d/kubernetes.repo
[kubernetes]
name=Kubernetes
baseurl=https://mirrors.aliyun.com/kubernetes/yum/
  repos/kubernetes-el7-x86_64/
enabled=1
gpgcheck=1
repo_gpgcheck=1
gpgkey=https://mirrors.aliyun.com/kubernetes/yum/
  doc/yum-key.gpg https://mirrors.aliyun.com/kubernetes/
  yum/doc/rpm-package-key.gpg
EOF
# install kubeadm、kubectl、kubelet
yum install -y kubectl-1.16.0-0 kubeadm-1.16.0-0
  kubelet-1.16.0-0
# start kubelet service
systemctl enable kubelet && systemctl start kubelet
```

After completing the installation of the three Kubernetes components, enter the following commands to configure the node:

```
mkdir -p $HOME/.kube
sudo cp -i /etc/kubernetes/admin.conf $HOME/.kube/
  config
sudo chown $(id -u):$(id -g) $HOME/.kube/config
```

Once the initialization of the master node configuration is complete, a token will be returned, which is necessary to add worker nodes to the cluster. This token should be stored securely. You can retrieve it later using the following command:

```
kubeadm token create --print-join-command
```

With the master node installation complete, you can use the command kubectl get nodes to check the status of the nodes. At this stage, the master node will be in the "NotReady" state.

......

Next, install the worker nodes. Like the master node, you need to install the kubeadm, kubelet, and kubectl components. The installation process is the same as for the master node. After starting the services, use the following command to join the cluster:

```
# join in clusters
kubeadm  join  192.168.10.104:6443  --token  ncfrid.
  7ap0xiseuf97gikl
   --discovery-token-ca-cert-hash sha256:47783e9851a1a
    517647f1986225f-104e81dbfd8fb256ae55ef6d68ce933
    4c6a2
```

Once the worker node successfully joins the cluster, you can verify it on the master node using the kubectl get nodes command. The output will show the worker nodes in the "Ready" state:

```
NAMESTATUSAGE
node1 Ready 1d
node2 Ready 1d
```

Exercises

(1) Describe how Docker containers are layered.
(2) What are the common commands for Dockerfile files?
(3) Why do you need Kubernetes when you have containers?

Chapter 8

Hadoop: A Distributed Big Data Development Platform

Hadoop is developed using the Java programming language and is an open-source implementation of Google's core technologies such as MapReduce, GFS (Google File System), and Bigtable. Supported by the Apache Software Foundation, it comprises the Hadoop Distributed File System (HDFS) and MapReduce (an open-source implementation of Google MapReduce) as its core components, along with several other subprojects that support Hadoop. This distributed big data development platform is primarily used for the efficient storage, management, and analysis of massive amounts of data (with PB-level data being the threshold of big data). This chapter primarily explains the architecture of Hadoop, the functions and working principles of its components, its basic composition, as well as the installation and basic usage of Hadoop.

8.1. Introduction to Hadoop

Hadoop is an open-source computing framework under the Apache Software Foundation, characterized by high reliability and good scalability. It can be deployed on a large number of low-cost hardware devices (PCs) to provide underlying support for distributed computing tasks. This section mainly introduces what Hadoop is, the architecture of Hadoop, and the basic functions of its components. It also explains the differences between Hadoop distributed development and general distributed development models.

8.1.1. *Hadoop and distributed development technology*

Today, most computer software runs on distributed systems, where the interactive interface, business processes, and data resources are stored on loosely coupled computing nodes and layered services, connected through a network. Distributed development technology has become the core technology for establishing application frameworks and software components, demonstrating strong vitality in developing large-scale distributed application systems. The concept of software development engineering in distributed systems has continuously evolved, with recent trends such as microservices and service meshes fundamentally expanding the divide-and-conquer philosophy advocated by distributed development.

Different distributed systems or development platforms operate at different levels and perform various functions, such as distributed operating systems, distributed programming languages and their compilation (interpretation) systems, Distributed File Systems (DFS), and distributed database systems. Developing a distributed system involves many tasks, so distributed development involves selecting specific distributed software systems or platforms according to user needs and then further developing or utilizing distributed applications based on these systems or platforms.

Hadoop is a type of distributed development technology that implements the functionalities of a DFS and parts of a distributed database. For instance, a single-node system with only 500 GB cannot process PB-level data at once. To solve this problem, large-scale datasets need to be stored across multiple different node systems, creating a file system with resources spread across multiple networked nodes, known as a DFS. Hadoop, with its core components, HDFS and MapReduce, serves as a foundational distributed system infrastructure mainly used for the efficient storage, management, and analysis of massive data. The high fault tolerance and scalability of HDFS in Hadoop allow users to deploy Hadoop on low-cost hardware, forming a DFS. The parallel programming framework MapReduce in Hadoop enables users to develop distributed parallel programs without understanding the underlying details of distributed systems, leveraging the power of clusters for high-speed computation and storage and allowing software developers to conduct corresponding distributed parallel software development through Hadoop.

To conduct distributed development using Hadoop, it is essential to understand its application characteristics. Hadoop has the capability to handle large-scale distributed data, and all data processing tasks are batch-processed, requiring all data to be local. Task processing involves high latency. Although the MapReduce processing is stream-based, it does not handle real-time data, indicating that Hadoop is not advantageous for real-time data processing.

Hadoop originated from Nutch, an open-source search engine implemented in Java and developed in 2002 by the Yahoo! team led by Doug Cutting. In 2003, Google published a paper on GFS, a distributed storage file system, at the Symposium on Operating Systems Principles (SOSP). In 2004, Google published a paper on MapReduce, a distributed processing technology, at the Operating Systems Design and Implementation (OSDI) conference. Cutting realized that GFS could address the storage needs of extremely large files generated during web crawling and indexing, while the MapReduce framework could be used to handle the indexing of massive web pages. However, Google only provided the concepts without open-sourcing the code. Therefore, in 2004, the Nutch project team replicated and rebuilt these two systems, forming Hadoop, which became a scalable technology applicable to web data processing. Figure 8.1 shows the Hadoop logo.

8.1.2. *Hadoop architecture*

Hadoop implements a system framework for distributed parallel processing of big data, employing a data-parallel processing approach. It combines the MapReduce computing framework, which performs data analysis, with the HDFS, which handles data storage. Hadoop automatically divides applications into many small work units and distributes these units across the corresponding nodes in a cluster, while HDFS manages data storage on each node, enabling high-throughput data read and write operations. The architecture of Hadoop is illustrated in Figure 8.2.

Fig. 8.1.   Hadoop logo.

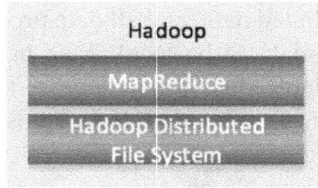

Fig. 8.2.   Hadoop architecture.

HDFS is Hadoop's distributed file storage system. From a user's perspective, HDFS functions similarly to other DFS, offering capabilities such as creating, deleting, moving, and renaming files. However, HDFS is designed for storing large amounts of data and is inherently distributed, which means its features are tailored to big data and distributed environments. To meet the demands of big data processing, Hadoop optimizes access to very large files, ensures that read operations far exceed write operations, and addresses issues such as node failures within a cluster.

MapReduce is a distributed computing framework and a fundamental component of Hadoop. It consists of two phases: the Map phase and the Reduce phase. MapReduce is a programming model that supports distributed parallel computation of datasets on the scale of PBs using inexpensive computer clusters. It comprises Map functions and Reduce functions, which handle task decomposition and result aggregation, respectively. MapReduce is used for batch processing rather than real-time querying, making it particularly unsuitable for interactive applications. It greatly simplifies the process for programmers to run their programs on a distributed system without needing expertise in distributed parallel programming.

Notably, since Hadoop's official release of version 1.0 in 2011, its basic architecture has remained stable. The 2.0 version, released in 2012, made significant updates to the original architecture by introducing enhancements such as HDFS Federation and YARN. The 3.0 version, released in 2017, further optimized the system to improve usability. Subsequent discussions on HDFS and MapReduce will also incorporate these new features.

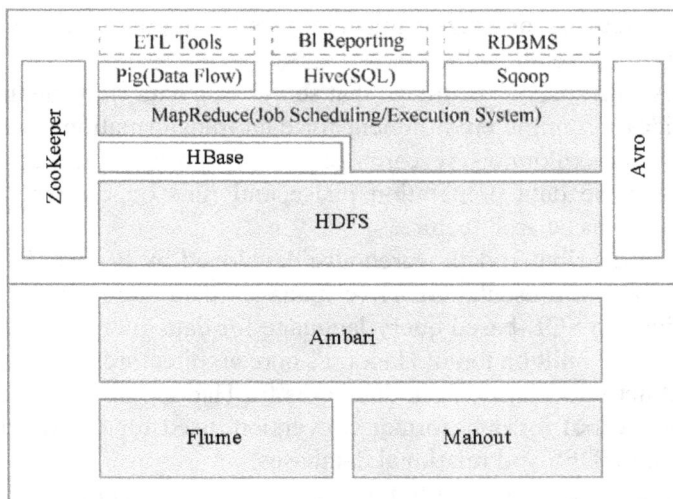

Fig. 8.3. Hadoop ecosystem diagram.

Currently, Hadoop has evolved into a collection of numerous projects, forming an ecosystem centered around Hadoop (the Hadoop Ecosystem), as shown in Figure 8.3. This ecosystem provides complementary services or higher-level services on top of the core layers, making Hadoop applications more convenient and efficient.

Here is an introduction to the Hadoop ecosystem:

- *YARN* (*Yet Another Resource Negotiator*): YARN is the resource management system for Hadoop clusters. Hadoop 2.0 includes a complete redesign of the MapReduce framework, referred to as MRv2 or YARN.
- *ETL Tools*: These are crucial for building data warehouses and consist of various data warehouse collection tools.
- *BI Reporting* (*Business Intelligence Reporting*): This provides comprehensive reporting, data analysis, and data integration capabilities.
- *RDBMS* (*Relational Database Management System*): An RDBMS stores data in databases known as tables, which are collections of

related records organized into rows and columns, forming a two-dimensional relational table.

- *Pig*: A data analysis language that serves as a data flow language and provides a runtime environment for data transformation (using pipelines) and exploratory research (e.g., rapid prototyping). Pig is suitable for the data preparation phase and runs on clusters built on Hadoop's basic architecture.

- *Hive*: A distributed data warehouse developed by Facebook, specializing in data presentation. Hive manages data stored in HDFS and provides an SQL-based query language for data querying. Both Hive and Pig are built on top of Hadoop's core architecture and can be used to extract information from databases for Hadoop processing.

- *Sqoop*: A tool for data format conversion, used for transferring data between HDFS and relational databases.

- *HBase*: A distributed column-oriented database similar to Google Bigtable. HBase supports parallel computing with MapReduce and point queries (i.e., random reads). It is a Java-based product, with its C++ counterpart being the open-source project Hypertable, also an Apache project.

- *Avro*: A new data serialization and transport format designed to replace the original Inter-Process Communication (IPC) mechanism in Hadoop's basic architecture.

- *Zookeeper*: A coordination service used for building distributed applications. It is a distributed locking facility that provides functionality similar to Google Chubby, mainly addressing distributed consistency issues. Zookeeper is developed by Facebook and is based on HBase and HDFS.

- *Ambari*: Ambari aims to add core functionalities such as monitoring and management to Hadoop. It assists system administrators in deploying and configuring Hadoop, upgrading clusters, and providing monitoring services.

- *Flume*: A highly available, reliable, and distributed tool for massive log collection provided by Cloudera. Flume supports customization of various data sources in log systems for data collection and offers capabilities for simple data processing and writing to various customizable data sinks.

- *Mahout*: A distributed framework for machine learning and data mining, distinct from other open-source data mining software, as it is based on Hadoop. Mahout implements some data mining algorithms using MapReduce, addressing the challenges of parallel mining.

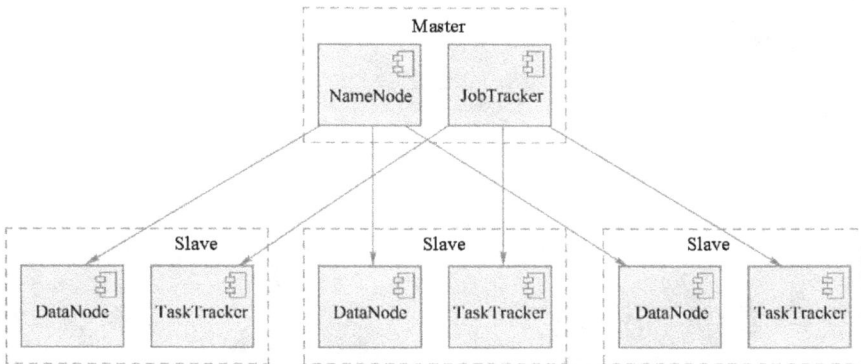

Fig. 8.4. Logical architecture of a Hadoop cluster.

8.1.3. *Hadoop cluster architecture*

The logical architecture of a Hadoop cluster follows a master-slave structure (Master/Slave architecture). The Master Node includes the NameNode (management node) and JobTracker (job server); the slave nodes include DataNodes (data nodes) and TaskTrackers (task servers). A Hadoop cluster comprises multiple DataNodes and TaskTrackers. The logical architecture of a Hadoop cluster is illustrated in Figure 8.4.

The master-slave structure of NameNode and DataNodes works together to complete HDFS functionality. The NameNode is responsible for receiving user operation requests (such as creating folders, deleting, moving, and traversing), maintaining the directory structure of the file system, and managing the relationships between files and blocks, as well as blocks and DataNodes. The DataNodes are responsible for storing files, with files divided into blocks and stored on disks (to ensure data safety, files have multiple replicas).

The master-slave structure of JobTracker and TaskTracker works together to complete MapReduce functionality. The JobTracker is responsible for receiving computation tasks submitted by clients, distributing these tasks to TaskTrackers for execution, and monitoring the execution status of TaskTrackers. The TaskTrackers are responsible for executing the computation tasks assigned by the JobTracker.

The physical distribution of a Hadoop cluster is shown in Figure 8.5. Clients interact with the backend Hadoop cluster through multiple switches (which perform tasks such as filtering, learning, and forwarding). A Hadoop cluster includes at least one NameNode and one

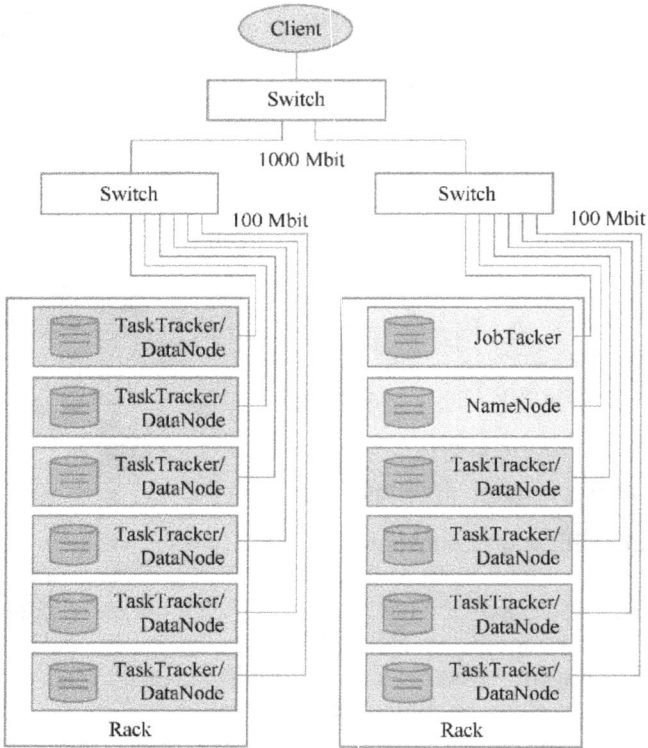

Fig. 8.5.   Physical distribution diagram of a Hadoop cluster.

JobTracker, multiple DataNodes, and multiple TaskTrackers. The physical distribution of a Hadoop cluster can consist of multiple racks. The cluster illustrated in Figure 8.5 consists of two racks.

The Master Node runs the JobTracker and NameNode, while the Slave node runs the TaskTracker and DataNode. The physical structure of a single-node Hadoop setup is illustrated in Figure 8.6. The physical structure of the Master Node from bottom to top includes a server, a Linux operating system (Operating System, OS), a Java Virtual Machine (JVM), Hadoop utilities, JobTracker, Secondary NameNode, NameNode, and a browser. The physical structure of the Slave node (Slave Node) from bottom to top includes a server, Linux OS, JVM, TaskTracker, and DataNode.

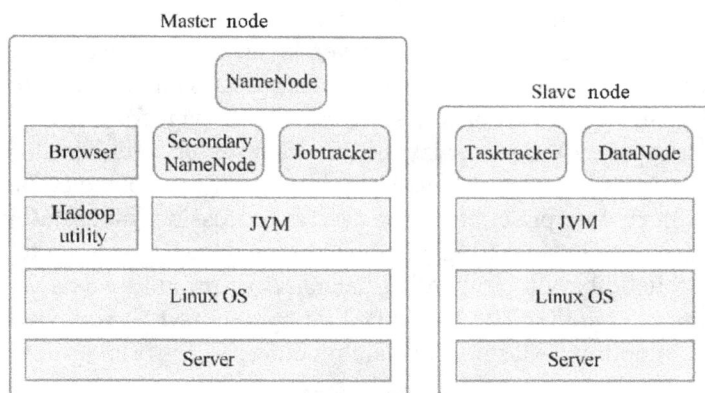

Fig. 8.6. Physical structure diagram of a single-node Hadoop setup.

8.2. Distributed File System HDFS

HDFS is a DFS designed to run on commodity hardware, featuring high fault tolerance and providing high throughput, making it suitable for applications with large datasets.

8.2.1. *Overview of DFS*

(1) *Concept of DFS*: To meet modern file storage requirements—such as large capacity, high reliability, high availability, high performance, dynamic scalability, and easy maintenance—DFS was introduced. As the name suggests, a DFS combines the characteristics of distributed systems and file systems. DFS makes files spread across multiple nodes appear as if they are located in a single position on the network, facilitating dynamic expansion and maintenance.

A DFS has two main aspects. From the perspective of the file system's users, it is a standard file system that provides a series of APIs to create, move, delete, and read/write files or directories. Internally, a DFS does not manage local disk storage like traditional file systems. Instead, the contents and directory structures of files are transmitted over the network to remote systems. This means the physical storage resources managed by a DFS are not necessarily directly connected to local nodes but are connected via a computer network.

Figure 8.7 illustrates the structure of a DFS, which includes three parts: clients, the application server layer, and the database service layer. Clients using cache listen to requests sent to the application server and store responses, such as HTML pages and images. The typical application server layer is the Web server layer, also known as the business logic layer. This layer consists of distributed components that meet enterprise business needs, processing input/output data according to business logic and accessing database servers to ensure data reliability before updating databases or providing data to users.

(2) *The Earliest DFS*: The Network File System (NFS) was the earliest DFS, adopting a Master/Slave architecture. NFS's primary function is to allow different machines and operating systems to share and manage files over a network. NFS enables the mounting of remote host directories, making them accessible as if they were local directories, effectively functioning as a file server. Figure 8.8 illustrates the architecture and workflow of the NFS file system.

The operation process of the NFS file system is as follows:

(1) Provide a shared directory, for example: /home/sharefile/.
(2) Collect file information from all clients.
(3) Each client mounts the shared directory to a local directory, enabling file sharing.
(4) Each client can access files shared by all other clients.

From the above process, it can be seen that NFS employs a Master/Slave architecture to achieve distributed file storage, but all file transfers between nodes must go through the Master Node.

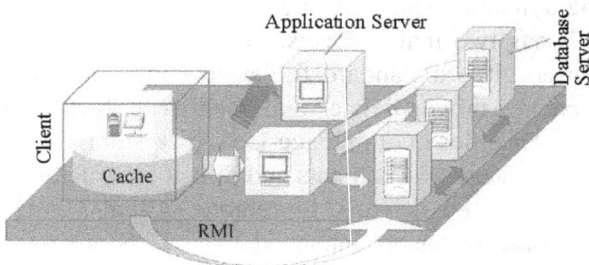

Fig. 8.7. Structure of a DFS.

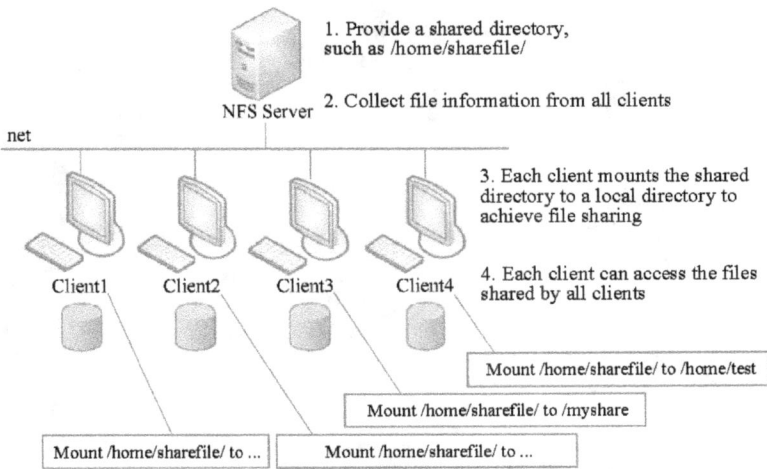

Fig. 8.8. Architecture and workflow of the earliest NFS file system.

(3) *Optimization Approaches for DFS in Big Data Environments*: In the era of big data, it is imperative for DFS to store and manage petabyte-scale data, handle unstructured data, and emphasize throughput in data processing. The read operations for big data significantly outweigh write operations. DFS architectures operate within cluster environments, where node failures are common due to the high likelihood of faults among the nodes. Nevertheless, a DFS must ensure that the failure of any single node does not affect the service provided to users. Therefore, technical optimizations of the DFS are required. The solution involves first dividing large files into blocks, which are then stored on different data nodes. Since data are stored on clusters of inexpensive and unreliable nodes, the number of data replicas should not be fewer than two. This ensures that file blocks remain intact even if a single data node fails, with subsequent processes ensuring sufficient replication. During data read and write operations, concurrent read and write are implemented on different data nodes, employing a write-once-read-many access model.

These strategies were initially proposed by Google and implemented in the GFS. For instance, when storing a large file, it is first divided into chunks of 64 MB, as illustrated in Figure 8.9. Each chunk is then assigned to server nodes in the cluster, which inform compute nodes where to store these chunks. The compute nodes then directly

File

1 | 2 | 3

Compute Node

Metadata
Server

LAN

1 | 3

2 | 3

1 | 2

X86
DataNode

X86
DataNode

X86
DataNode

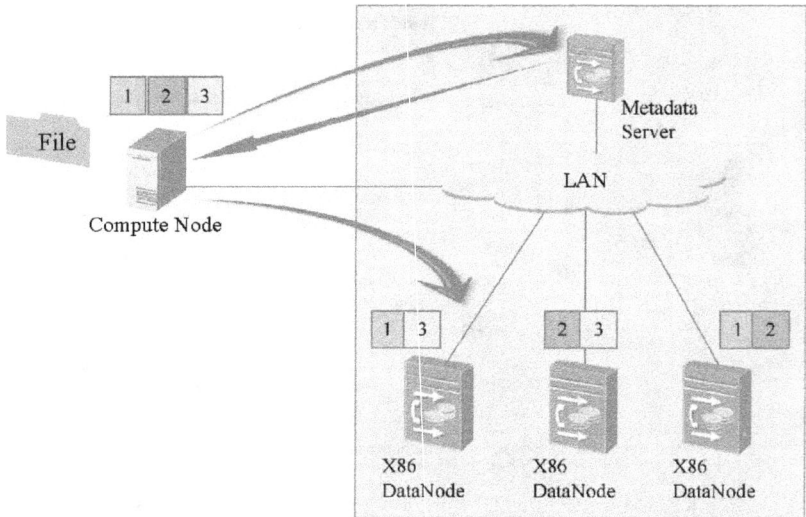

Fig. 8.9.   Storage approach of Google DFS.

store the data on DataNode nodes, bypassing the server nodes for data transfer. This approach alleviates the data transfer bottleneck at server nodes. Server nodes primarily handle metadata, which includes information about data properties and functions as an electronic directory.

The distinctions between DFS and NFS in a big data environment are evident: In a big data environment, DFS enables data transfer without routing through the NameNode, achieving file block division and multiple access points, while also providing computational capabilities. Conversely, in NFS, data transfer must go through the NFS Server, files are not divided into blocks, and it provides destination sharing but lacks computational functionality.

8.2.2. HDFS architecture and Read/Write process

(1) *HDFS Architecture*: HDFS is characterized by a typical Master/Slave architecture. The Master Node (NameNode), also known as the Metadata Node, acts as the manager within the DFS, storing the system's metadata. This metadata includes the management node of the file system (NameNode), access control information, the current location of blocks, cluster configuration, and other information. The Slave

nodes, also known as DataNodes, provide the physical support for the actual file data. A Hadoop cluster contains numerous DataNodes that respond to client read and write requests and also execute commands from the Metadata Node for creating, deleting, moving, and replicating file blocks. The architecture of HDFS is illustrated in Figure 8.10.

From Figure 8.10, it can be seen that clients can read data blocks from multiple DataNodes through the Metadata Node. The collection of file metadata information is spontaneously submitted by each DataNode to the Metadata Node, which stores the basic information of the files. When there are changes in the file information on a DataNode, the updated file information is transmitted to the Metadata Node. The Metadata Node uses this metadata information to locate DataNodes for read operations. This critical information is usually backed up and stored on a Secondary Metadata Node. For writing files, it is necessary to know the metadata of each node, such as which blocks are free, the locations of free blocks, proximity to DataNodes, and the number of replicas before writing. In scenarios with at least two racks, data are generally written not only to several nodes within the local rack but also to a node in another rack, implementing what is known as "rack awareness". In Figure 8.10, Rack1 and Rack2 represent two such racks.

In HDFS, DataNodes report the stored file block information to the MetadataNode using a heartbeat mechanism. This mechanism involves sending regular reports to the NameNode about the block mapping status and metadata information. If these reports do not reach the MetadataNode within a specified time, the MetadataNode

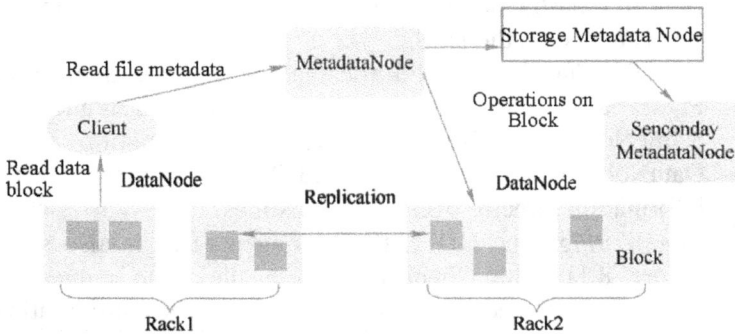

Fig. 8.10.   HDFS system architecture diagram.

will consider the node to be uncommunicative. If there is no heartbeat message for an extended period, the node is marked as dead, and the MetadataNode will cease to monitor it unless the node manually contacts the NameNode upon recovery. This process is also referred to as block operations.

Files on HDFS are divided into multiple 64 MB chunks, which serve as independent storage units. Data that do not fill an entire chunk will not occupy the whole block space, allowing this space to be shared with other data. The abstraction of blocks in a DFS is a highly effective design, bringing several advantages. For instance, HDFS can store a file larger than any single disk capacity within the cluster because of its block storage approach.

By replicating blocks, HDFS enhances the fault tolerance and reliability of the file system. Block redundancy is maintained by replicating blocks to several other nodes (with a default of three replicas). If a block becomes corrupted, its replica can be read from another node and then redundantly backed up to additional nodes. This process is automatically managed by HDFS and is transparent to the user.

HDFS is not suitable for storing small files. If the files are relatively small, it is not recommended to use HDFS. Additionally, HDFS is not recommended for scenarios involving a large number of random read operations. It is also important to note that HDFS does not support file modifications.

(2) *HDFS Read/Write Process*: The Master Node (NameNode) manages all the metadata of the file system, providing a hierarchical structure of metadata for all files and directories. It controls client file operations and manages the allocation of storage tasks, determining the mapping of data blocks to DataNodes. The Client is the application that needs to access the DFS.

- *File Reading*: The HDFS file reading process is illustrated in Figure 8.11. The Client initiates a file read request to the NameNode. The NameNode returns information about the DataNodes where the file is stored. The Client then reads the file information directly from the DataNodes.

- *File Writing*: The HDFS file writing process is also shown in Figure 8.11. The Client initiates a file write request to the NameNode. Based on the file size and block configuration, the NameNode returns information about the relevant DataNodes it manages. The Client divides the file into multiple blocks and,

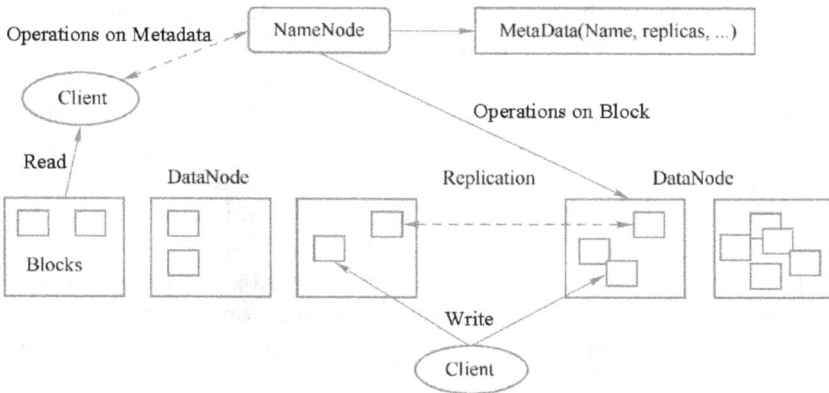

Fig. 8.11. HDFS file reading operation diagram.

using the address information of the DataNodes, writes these blocks directly to each DataNode in sequence, bypassing the NameNode.

(3) *Issues and Developments*: With the rapid development of the Internet, the volume of data that Hadoop needs to process has grown geometrically. While this architecture is well suited for handling massive data storage, it encounters significant issues when dealing with a large number of files. This is because the metadata for all data in the cluster is maintained by the NameNode to ensure efficient access. Each file, directory, and data block in HDFS is recorded in the NameNode's memory, with each data block's information occupying approximately 150 bytes. The vast amount of single-point data thus becomes a bottleneck for Hadoop's stability and scalability.

To address the horizontal scalability issue of HDFS, the community introduced HDFS Federation starting from Apache Hadoop version 0.23.0. HDFS Federation allows multiple NameNodes/ Namespaces to coexist within the HDFS cluster, with each Namespace being independent of the others. Each Namespace contains multiple NameNodes, one of which is the primary and the others are backups. These Namespaces collectively manage the data across the cluster, with each Namespace managing only a portion of the data, operating independently of the others. DataNodes in the cluster register with all NameNodes, periodically sending heartbeats and block information, and execute commands sent by the NameNodes. The NameNodes in the cluster share the storage resources of all DataNodes.

However, while HDFS Federation addresses the issue of horizontal scalability, it introduces new challenges. With multiple Namespaces in the cluster, the client needs to know which Namespace contains the data it wants to query. In version 2.0, the community introduced the View File System to address this issue, but this solution places the resolution responsibility on the client and involves significant migration costs for users upgrading from version 1.0. To mitigate this, the community introduced a router-based Federation scheme in Hadoop versions 2.9.0 and 3.0.0. Due to space constraints, this solution will not be discussed further here; interested readers can refer to the official HDFS documentation for more details.

8.3. Distributed Computing Framework MapReduce

MapReduce is a framework designed for performing reliable, fault-tolerant distributed computation on large-scale commercial hardware clusters (comprising thousands of nodes) across massive datasets (ranging in the terabytes). It is also a classic parallel computing model. The fundamental principle of MapReduce involves dividing a complex problem (dataset) into several simpler subproblems (data chunks) for processing (Map function). The results of these subproblems are then merged (Reduce function) to obtain the solution to the original problem (result), as illustrated in Figure 8.12. Additionally, the MapReduce model is well-suited for processing large files, but its efficiency in handling numerous small files is not as high, similar to HDFS.

8.3.1. *MapReduce programming model*

(1) *Introduction to the MapReduce Programming Model*: MapReduce is both a concept and a programming model. For Hadoop, MapReduce is a distributed computing framework and a fundamental component of the system. Once the Hadoop cluster is configured, MapReduce is included by default.

The MapReduce programming model primarily consists of two abstract classes: the Mapper class and the Reducer class. The Mapper is used to process the split raw data, while the Reducer aggregates the

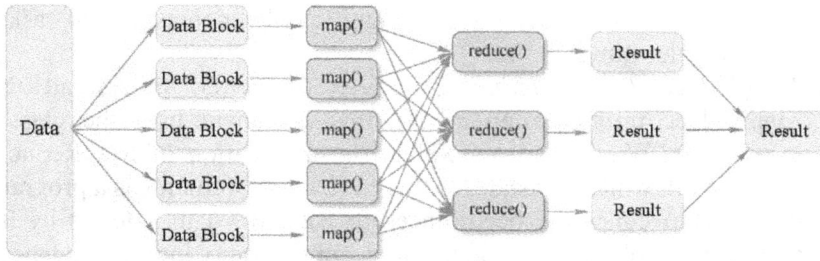

Fig. 8.12.   Basic principles of MapReduce diagram.

results from the Mapper to produce the final output. For software developers, writing a MapReduce program involves implementing the Map and Reduce functions, which is as straightforward as writing process functions.

In terms of data format, the Mapper accepts data in the form of <key, value> pairs and produces a series of output records in the same <key, value> format. These outputs are then processed to form intermediate results in the form of <key, {value list}>. The intermediate results produced by the Mapper are subsequently passed to the Reducer as input. The Reducer processes the {value list} for each key and generates the final <key, value> results, which are then written to HDFS. Based on its working principles, the MapReduce programming model can be categorized into two types: the simple MapReduce model and the complex MapReduce model.

- *Simple MapReduce Model*: For certain tasks, the Reduce process may not be necessary. For instance, if the only requirement is to perform a simple format transformation on each line of text data, then processing by the Mapper alone suffices. In such cases, the MapReduce model can be simplified to include only the Mapper process, where the data produced by the Mapper is directly written to HDFS.
- *Complex MapReduce Model*: For most tasks, the Reduce process is required. Due to the substantial workload, multiple Reducers are initiated (defaulting to one, but the number can be adjusted by the user based on the task's volume) to perform the aggregation. Using a single Reducer to compute results from all Mappers can lead to excessive load on that Reducer, becoming a

performance bottleneck and significantly increasing the task's execution time.

(2) *MapReduce Programming Example*: To better understand the MapReduce programming model, this document uses the classic example of WordCount to explain the application of MapReduce. WordCount, which translates to "word frequency count", is a program designed to count the occurrences of each word in a text file. Its approach involves splitting the text content into individual words using "space" as the delimiter, without validating whether these words are actual words. The input can consist of multiple files, but there is only one output.

WordCount serves as an introductory program for learning Hadoop. It is one of the simplest and most illustrative examples of the MapReduce concept.

You can start by creating two small files, as illustrated in the following:

```
File: text1.txt File: text2.txt
hadoop is very good hadoop is easy to learn
mapreduce is very good mapreduce is easy to learn
```

Then, you can upload these two files to HDFS and process them using WordCount (refer to subsequent chapters for the operation process). The final results will be stored in the specified output directory. Opening the result file will show the following content:

```
easy 2
good 2
hadoop 2
is 4
learn 2
mapreduce 2
to 2
very 2
```

From the results above, it can be observed that each line contains two values separated by a tab character. The first value is the key, which represents the word identified by WordCount, and the second value is the value, indicating the frequency of each word. Observant readers might notice that the results are sorted in ascending order by key, which reflects the sorting performed during the MapReduce process.

The pseudocode for implementing WordCount is as follows.

```
mapper(String key, String value)   // key: offset,
    value: string content
{
    words = SplitInTokens(value);   // Split the
                                    string into tokens
    for each word w in words        // For each word
                                    in the string
        Emit(w, 1);                 // Emit the word
                                    with a count of 1
}
reducer(string key, value_list)     // key: word,
                                    value_list: list
                                    of values
{
    int sum = 0;
    for each value in value_list    // For each value
                                    in the list
        sum += value;               // Add to the
                                    variable sum
    Emit(key, sum);                 // Emit the key
                                    with the total
                                    count
}
```

The pseudocode above illustrates the Mapper and Reducer processes for WordCount. In actual implementation, multiple Mapper and Reducer instances are typically used, running on different nodes. Initially, each Mapper processes its input to tokenize the text and outputs intermediate results in the form of <word, 1>, which are stored on the local disk of each node. Subsequently, Reducers aggregate these results, with different Reducers handling different segments of the data, computing the total count for each word, and finally outputting the results in the form of <word, counts> to HDFS.

From this example, it is evident that the MapReduce programming model has limitations on the types of problems it can handle. MapReduce is suitable for scenarios where large problems are divided into smaller, independent subproblems. For instance, in this example, calculating word counts in text1 does not affect the calculation for text2, and vice versa.

8.3.2. *MapReduce data flow*

From the MapReduce programming model, it is apparent that data flow in various forms between different nodes. It is processed at each node and then transformed into a different form as it moves to the next node, ultimately leading to the final result. Therefore, understanding the data flow in and out of each node is crucial for system development.

Mappers handle data in the <key, value> format and cannot directly process file streams. So, how do Mapper data get sourced? How are data produced by multiple Mappers allocated to multiple Reducers? These operations are managed by Hadoop's core APIs, such as InputFormat, Partitioner, and OutputFormat. These APIs function similarly to Mappers and Reducers, operating at the same level but performing different tasks. They come with many default implementations that meet most user needs. However, if the default implementations do not satisfy specific requirements, users can extend and override these basic classes to implement custom processing.

Using the WordCount example, the MapReduce data flow process is illustrated in Figure 8.13.

(1) *Splitting and Formatting Data Source (InputFormat)*: InputFormat has two main tasks. First, it splits the source files and determines the number of Mappers required; second, it formats each split into a <key, value> data stream and passes it to the Mapper. In the diagram, the source file is first split into four partitions, and the number of Mappers is determined (four). Each partition is then formatted into a <key, value> data stream and passed to the Map() function.

(2) *Map Process*: The Mapper receives data in <key, value> format and processes it into another <key, value> format, with the specific processing defined by the user. In WordCount, the Mapper parses the incoming key values, using "space" as the delimiter. When encountering a "space", the accumulated string before the delimiter is output as the key, with a value of 1 for each occurrence, resulting in <word, 1> pairs.

(3) *Combiner Process*: Each map() operation may produce a large amount of local output. The Combiner() function performs a local merge of

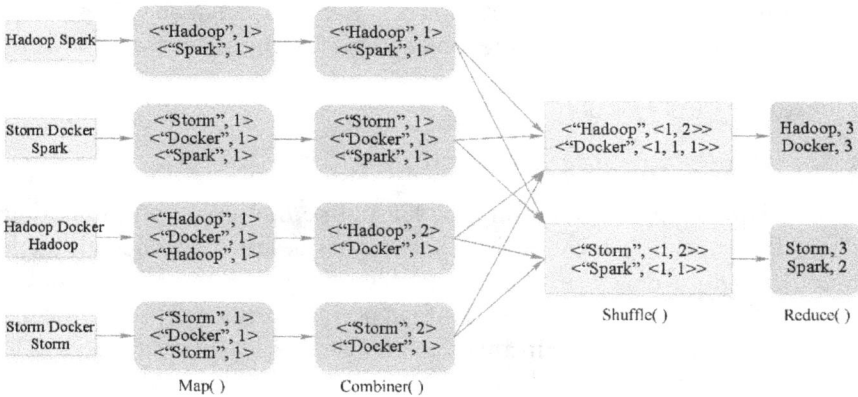

Fig. 8.13.   WordCount processing flow.

the map() output to reduce data transfer between Map and Reduce nodes and improve network I/O performance. This is one of the optimizations in MapReduce. For example, in WordCount, the map() output is merged before being passed to the Combiner().

(4) *Shuffle Process*: The Shuffle process refers to the series of steps that transform the direct output from Mappers into the final input data for Reducers. This is a core component of MapReduce.

The Shuffle process can be divided into two phases: Mapper-side Shuffle and Reducer-side Shuffle. Data produced by Mappers are not immediately written to disk but are first stored in memory. Once the memory data reach a threshold, they are written to a local disk and undergo sorting, combining, and partitioning. The sort operation arranges Mapper outputs by key value; the combine operation merges adjacent records with the same key value; and the partition operation distributes data evenly among multiple Reducers, directly affecting load balancing. The combine operation is optional and may not be used in all scenarios, but it is often employed to compact Mapper outputs.

Mappers and Reducers generally run on different nodes, with Reducers being fewer in number than Mappers. Consequently, Reducers must download and process results from multiple Mappers, which involves additional processing during the Reducer-side Shuffle.

(5) *Reduce Process*: The Reducer receives data in <key, {value list}> format and processes it to produce <key, value> format output. The output is directly written to HDFS, with the specific processing

defined by the user. In WordCount, the Reducer sums the values in the list for each key to obtain the total count of occurrences for each word and outputs this result.

8.3.3. *MapReduce task execution flow*

In the MapReduce programming model, unlike traditional programming where the execution location is fixed and tasks are performed through direct calls to a database, the MapReduce program dynamically flows across different nodes. To avoid confusion with traditional program concepts, we refer to a program that accomplishes a specific function as a "task". The MapReduce task flow encompasses the series of steps from task submission by the client to task completion. MRv2 represents the MapReduce task execution flow in Hadoop 2. In MRv2, the MapReduce runtime environment is provided by YARN, requiring collaboration between MapReduce services and YARN services. In the following, we first describe the basic components of MRv2 and YARN, followed by an overview of the MapReduce task execution flow:

(1) *Basic Components of MRv2*: MRv2 discards the JobTrack and TaskTrack from MRv1 (the MapReduce task execution process in Hadoop 1) in favor of a new MRAppMaster for managing individual tasks. MRAppMaster works in coordination with YARN's ResourceManager and NodeManager to schedule and control tasks, thus addressing the issue of excessive load on a single service (JobTrack in MRv1) responsible for managing and scheduling all tasks. The basic components of MRv2 are as follows:

- *Client*: The client submits tasks to the YARN cluster and is the sole communication channel between MapReduce users and the YARN cluster. It communicates with the ResourceManager via the ApplicationClientProtocol (an implementation of the RPC protocol) and can query task status or terminate tasks. The client can also communicate with the MRAppMaster through MRClientProtocol (another RPC implementation) to directly monitor and control jobs, thereby reducing the ResourceManager's load.
- *MRAppMaster*: MRAppMaster is an implementation of ApplicationMaster, responsible for monitoring and scheduling the

entire MR task workflow. Each MR task has a single MRAppMaster. The MRAppMaster manages tasks but does not handle resource allocation.

- *Map Task and Reduce Task*: These are instances of user-defined Map and Reduce functions. In MRv2, these tasks run within the resource constraints defined by YARN and are managed and scheduled collaboratively by MRAppMaster and NodeManager.

(2) *Basic Components of YARN*: Yet Another Resource Negotiator (YARN) is a resource management platform that monitors and schedules resources across the entire cluster. It is responsible for managing the execution of tasks and allocating resources within the cluster. The basic components of YARN are as follows:

- *Resource Manager* (*RM*): Running on the NameNode, the Resource Manager is the resource scheduler for the entire cluster and consists of two main components: the Resource Scheduler and the Applications Manager.
  - o *Resource Schedule*: When an application registers and requests resources, the ApplicationMaster submits resource requests to the Resource Scheduler. It allocates resources based on current availability and constraints and generates a container resource description.
  - o *Applications Manager*: This component manages all tasks running across the cluster, including handling application submissions, negotiating with the Resource Scheduler to start and monitor ApplicationMasters, and restarting ApplicationMasters on other nodes in the event of task failures.
- *NodeManager*: Running on the DataNodes, the NodeManager monitors and manages the computational resources of individual nodes. It periodically reports the resource usage of its node to the Resource Manager. When tasks are running on a node, the NodeManager is responsible for creating containers, monitoring their running state, and eventually destroying them.
- *ApplicationMaster* (*AM*): Responsible for scheduling and managing the lifecycle of a specific task flow, the ApplicationMaster handles task registration, resource requests, and communicates with the NodeManager to start and terminate tasks.

- *Container*: In the YARN architecture, a container is a representation of computational resources on a node. It encapsulates various dimensions of resources, including CPU, RAM, disk, and network. When the ApplicationMaster requests resources from the Resource Manager, the allocated resources are represented as containers. Map Tasks and Reduce Tasks must operate within the constraints defined by these container descriptions.

(3) *Task Flow*: In YARN, resource management is a collaborative effort between the Resource Manager and the NodeManager. The Resource Manager is responsible for resource allocation, while the NodeManager handles resource provisioning and isolation. Once the Resource Manager allocates resources on a specific NodeManager (referred to as "resource scheduling"), the NodeManager must provide the required resources for the task and ensure these resources are exclusive, thereby guaranteeing the foundational requirements for task execution (referred to as resource isolation).

In the YARN architecture, the MapReduce task execution process can be divided into two main parts. First, the client submits a task to the ResourceManager, which then notifies the corresponding NodeManager to launch the MRAppMaster. Second, once the MRAppMaster is successfully started, it is responsible for scheduling and managing the entire task execution until completion. The detailed steps of this process are illustrated in Figure 8.14.

(1) The client submits the task to the ResourceManager.
(2) The ResourceManager allocates the first container for the task and notifies the corresponding NodeManager to start the MRAppMaster.
(3) Upon receiving the command, the NodeManager allocates a container resource space and starts the MRAppMaster within the container.
(4) After the MRAppMaster starts, it first registers with the ResourceManager, allowing the user to monitor the task's running status directly through the MRAppMaster. Subsequently, the MRAppMaster schedules the task execution, repeating steps 5–8 until the task completes.
(5) The MRAppMaster requests the resources needed for task execution from the ResourceManager in a polling manner.

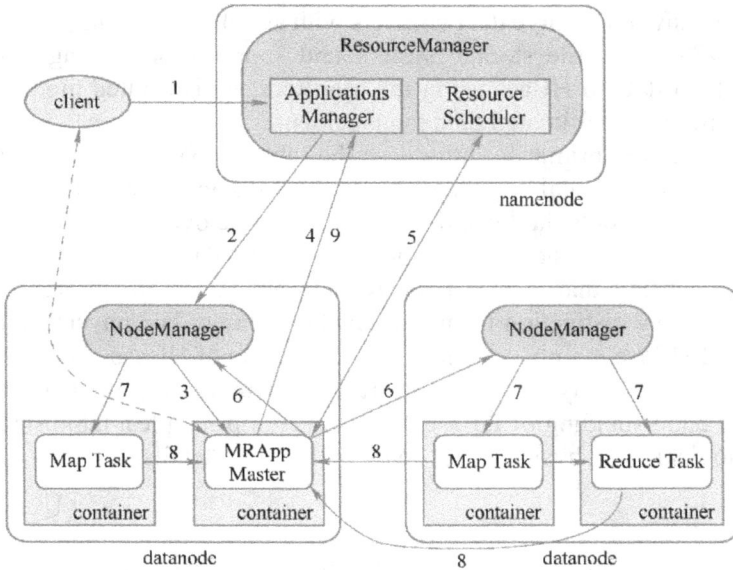

Fig. 8.14. Task execution flow of MapReduce in YARN.

(6) Once the ResourceManager allocates the resources, the MRAppMaster communicates with the corresponding NodeManager to allocate containers and start the relevant tasks (Map Task or Reduce Task).

(7) The NodeManager prepares the runtime environment and starts the tasks.

(8) The tasks run and periodically report their status and progress to the MRAppMaster via the RPC protocol. The MRAppMaster also monitors the tasks in real time, and if a task is found to be stuck or failed, it terminates and restarts the task.

(9) Upon task completion, the MRAppMaster communicates with the ResourceManager to deregister and shut down itself.

8.4. Column-Oriented Database HBase

HBase is a highly reliable, high-performance, column-oriented, scalable, real-time read-write distributed database system. It offers near-disk-limit write performance and excellent read capabilities, making it suitable for scenarios involving large data volumes with simple operations. HBase can use HDFS as its file storage system and supports processing massive amounts of data in HBase using the MapReduce distributed model, with Zookeeper for coordinated data management.

This section primarily introduces the table views of HBase (conceptual view and physical view) and its physical storage model. The conceptual view represents the logical view, showing the overall structure of the table, while the physical view represents the fundamental storage structure, illustrating that HBase table records are stored in different units, which is a key distinction from relational databases. The physical storage model of HBase mainly covers the basic services of HBase, data processing flow, and underlying data structures. Through this section, readers will gain an understanding of HBase's fundamental principles, table structure (both logical and physical), and some underlying details.

8.4.1. *Introduction to column-oriented database HBase*

HBase implements the principles of Google's BigTable, a sparse, column-oriented database, and is built on Hadoop's HDFS. This design leverages the high reliability and scalability of HDFS while incorporating the efficient data organization of BigTable. HBase provides an effective open-source solution for real-time response to massive data volumes.

HBase offers a Shell that is similar to relational databases like MySQL, which allows for control and manipulation of HBase tables and column families. The HBase Shell's help command provides detailed information on the commands supported by HBase, with specific usage methods available in the documentation.

Essentially, HBase operates as a map data structure, akin to arrays in PHP, dictionaries in Python, Hashes in Ruby, or Objects in JavaScript. Each row in HBase is a map, which can contain multiple maps (based on column families). Accessing data is similar to retrieving data from a map: by specifying a row key (to retrieve data from this map) and a key (column family name + qualifier) to obtain the data. In other words, a map is a "data structure consisting of keys and values, where each key is associated with a value".

HBase is a distributed database similar to BigTable, sharing most of its characteristics. It is a sparse, long-term, multi-dimensional, and sorted mapping table. The table's index is comprised of row keys, column keys, and timestamps. Each value is an uninterpreted character array, with data stored as strings and without explicit types. Users store data in tables, with each row having a sortable primary key and any number of columns. Due to its sparse storage nature, each row in the same table can have entirely different columns. Column names follow the format "<family>:<label>", composed of strings, and each table has a fixed set of families, defining the table structure. The structure can only be changed by altering the table schema, but label values can vary from row to row.

HBase stores data from the same column family in the same directory, and HBase's write operations are row-locked. Each row is an atomic element that can be individually locked.

All database updates in HBase are timestamped. Each update creates a new version, with HBase retaining a configurable number of versions. Clients can choose to retrieve the most recent version relative to a specific timestamp or fetch all versions at once.

HBase uses HDFS as its file storage system and supports processing massive amounts of data in HBase using the MapReduce distributed model, with Zookeeper for coordinated data management.

8.4.2. *Understanding the HBase table structure*

(1) *Conceptual View of HBase Tables*: HBase differs from conventional relational databases, which use two-dimensional tables for data storage, typically consisting of rows and columns. In relational databases, the attributes of columns must be defined beforehand, while rows can be dynamically expanded. In contrast, HBase tables are generally composed of Row Keys, Time Stamps, Column Families, and Rows, as illustrated in Table 8.1.

HBase tables can be conceptualized as large mappings where data can be located using the row key, or row key combined with a timestamp. Due to its sparse nature, some columns may be empty. Column families, on the other hand, must be predefined before use; they are similar to columns in a two-dimensional table but differ in that column families, columns, timestamps, and rows can be dynamically extended at runtime. This flexibility is a significant departure from traditional relational databases. The conceptual view of HBase is illustrated in Table 8.2.

- *Row Key*: The row key is the primary key used for retrieval in HBase, and each row can have only one row key. In HBase, tables

Table 8.1.   Composition of HBase tables.

Row Key	Time Stamp	Column Contents	Column Anchor	Column Mime
"com.cnn. www"	t9		anchor:cnnsi. com="CNN"	
	t8		anchor:my.look. ca="CNN.com"	
	t5	contents:html="…"		mine:type="text/ html"
	t4	contents:html="…"		
	t2	contents:html="…"		

Table 8.2.   Conceptual view of HBase tables.

Row Key	Time Stamp	Column *"contents:"*	Column *"anchor:"*		Column *"mime:"*
"com.cnn. www"	t9		"anchor:cnnsi.com"	"CNN"	
	t8		"anchor:my.look.ca"	"CNN.com"	
	t6	"<html>…"			"text/html"
	t5	"<html>…"			
	t3	"<html>…"			

can only be indexed by row keys. For instance, in Table 8.2, "com.cnn.www" is a row key. Row keys can be any string with a maximum length of 64 KB. In HBase, row keys are stored as byte arrays and do not have a specific data type, so sorting during storage does not consider data types. It is important to note that numerical values are not sorted in the way humans might expect. For example, sorting numbers 1–20 would result in the following sequence: 1, 10, 11, …, 19, 2, 20, 3, 4, 5, …, 9. Therefore, when designing row keys that include numerical values, it is advisable to use left-padding with zeros, such as 01, 02, 03, …, 19, 20.

- *Timestamp*: The timestamp is a time marker indicating when data were added, reflecting the freshness of the data. Each data entry defined by a row key and column qualifier is assigned a timestamp upon addition. The timestamp primarily serves to identify different versions of the same data. In Table 8.2, the data entry defined by the row key "com.cnn.www" and column "contents:html" has three versions, added at times t6, t5, and t3. To facilitate quicker retrieval of newer versions, data versions are stored in descending order based on their timestamps, so the latest data entry is found first when reading the storage file.

Timestamps are generally assigned automatically by HBase with the system time at the time of data writing but can also be explicitly specified by users. The data type for timestamps is a 64-bit integer, which allows for system time with millisecond precision.

While storing multiple versions of data has its advantages, managing excessive versions can be burdensome. Therefore, HBase provides two recycling mechanisms: one that retains a limited number of versions, discarding the oldest versions beyond this limit, and another that retains data versions only within a specified time range, discarding versions outside this range.

- *Column Family*: A column family is a collection of related columns where similar data are grouped together. In HBase, multiple column families can exist, but they must be defined before use. From the perspective of column families, HBase is structured, with column families serving a role similar to columns in relational databases. Column families are integral to the table and cannot be arbitrarily modified or deleted; such changes require taking the table offline.

  In terms of storage, HBase uses column families as storage units, meaning each column family is stored separately. This column-oriented storage design is a fundamental characteristic of HBase.

- *Column*: In HBase, columns do not exist as concrete entities but are instead virtual constructs composed of a column family name, a colon, and a qualifier. For instance, in Table 8.2, "anchor:cnnsi. com" and "mine:type" represent columns within their respective column families. Within a single column family, different qualifiers can be considered as distinct columns; for example, "anchor:cnnsi.com" and "anchor:my.look.ca" are two columns within the "anchor" column family. Columns do not need to be pre-defined; they are specified with qualifiers at the time of data insertion. From the perspective of columns, HBase is unstructured, as columns, like rows, can be dynamically expanded as needed.

- *Cell*: A cell in HBase is a unique table unit defined by the combination of a row key and column qualifier, containing a value and a timestamp indicating the version of that value. A cell's content is an indivisible byte array, and it represents the smallest operational unit within an HBase table. For example, in Table 8.2, each combination of a cell within a column family and its corresponding timestamp represents a cell. Notably, many cells in the table may be empty; these empty cells do not occupy storage space since they are not stored as data. This design contributes to HBase's logical sparsity.

- *Row*: Rows in an HBase table generally consist of a row key and one or more columns with associated values. Rows are stored and sorted lexicographically based on the row key. To optimize data locality and ensure similar data are stored close together, special attention should be given to row key design. For instance, if row keys are website domain names, using a reverse-order sorting method for keys like "org.apache.www", "org.apache.mail", and "org.apache.jira" will group all Apache-related web pages closely together in storage, avoiding dispersion due to variations in the initial segments of domain names (e.g., "www", "mail", and "jira").

- To better understand concepts like rows and column families, Figure 8.15 illustrates an example list snippet of HBase storing web pages. The row key is a reversed URL, such as "com.cnn. www". The "contents" column family stores web page content, while the "anchor" column family contains anchor text linking to the page. For example, CNN's homepage is referenced by "Sports Illustrated" (SI) and "MY-look", hence the row includes columns named "anchor:cnnsi.com" and "anchor:my.look.ca". Each anchor text has a single version identified by timestamps (e.g., t9 and t8), while the "contents" column has three versions, marked by timestamps t2, t4, and t5.

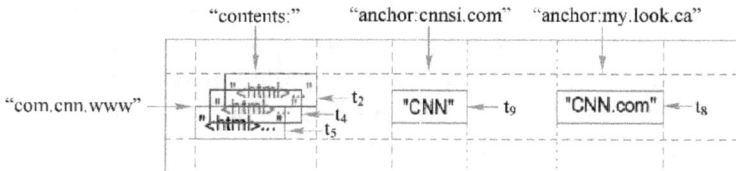

Fig. 8.15.   HBase storage example for web pages.

(2) *Physical View of HBase Tables*: In HBase, while the conceptual view of a table consists of sparse rows, many of which may lack complete column families, the physical storage is organized by column families. This means that a single row's data can be distributed across multiple physical storage units, with empty cells being discarded. For instance, a table with three column families, as shown in Table 8.2, will be divided into three storage units during physical storage, each corresponding to one of the column families. This is illustrated in the physical view in Table 8.3.

The advantage of organizing storage by column family is that new columns can be added to a column family at any time without prior declaration. Likewise, new column families can be introduced without necessitating modifications to the already stored physical units. This storage model renders HBase highly suitable for key-value queries.

From Table 8.3, it can be observed that the empty cells displayed in the conceptual view are not stored physically. Thus, if a query requests data for "contents:html" at timestamp "t8", no value will be returned, and similar requests will yield no result. However, if no timestamp is specified in the request, the latest version of the column's data will be returned. If no column is specified either, the latest value from each column will be returned. For a request to retrieve the value for row key "com.cnn.www", the returned values will include "contents:html" at "t5", "anchor:cnnsi. com" at "t9", "anchor:my.look.ca" at "t8", and "mine:type" at "t5".

As indicated, some columns that appear blank in the conceptual view are not actually stored. From the physical view, it is evident that querying

Table 8.3. Physical view of HBase table.

Row Key	Time Stamp	Column Contents
"com.cnn.www"	t5	contents:html="..."
"com.cnn.www"	t4	contents:html="..."
"com.cnn.www"	t2	contents:html="..."
Row Key	Time Stamp	Column anchors
"com.cnn.www"	t9	anchor:cnnsi.com="CNN"
"com.cnn.www"	t8	anchor:my.look.ca="CNN.com"
Row Key	Time Stamp	Column mime
"com.cnn.www"	t5	mine:type="text/html"

these blank cells will return null values. If no timestamp is provided during the query, the most recent version of the data will be returned, as the data are stored in timestamp order.

8.5. Setting Up a Hadoop Development Environment

This section describes how to set up a Hadoop cluster environment using four Linux virtual machines. One virtual machine will serve as the NameNode (Master Node), and the remaining three virtual machines will serve as DataNodes (Slave nodes). The details for the virtual machines are as follows:

- *Virtual Machine* 1: Hostname: vm1, IP: 192.168.122.101, used as the NameNode.
- *Virtual Machine* 2: Hostname: vm2, IP: 192.168.122.102, used as a DataNode.
- *Virtual Machine* 3: Hostname: vm3, IP: 192.168.122.103, used as a DataNode.
- *Virtual Machine* 4: Hostname: vm4, IP: 192.168.122.104, used as a DataNode.

Hadoop is a Java-based project, so JDK support is required. Installation and configuration on each node in the Hadoop cluster are identical. Therefore, you can start by installing and configuring Hadoop on one virtual machine and then replicate this configuration on the other nodes. This process will complete the setup of the Hadoop development environment. This section guides you through the setup of the Hadoop development environment and demonstrates running a test program, WordCount, to enhance your understanding of Hadoop through practical experience.

8.5.1. *Relevant preparations*

The preparatory steps before setting up the Hadoop development environment are as follows:

(1) *Prepare the Virtual Machine Operating System*: Begin by setting up four virtual machines with an operating system. For this section, all

virtual machines use CentOS 6.5 (64-bit). First, install the operating system on one virtual machine and then clone this machine to create the other three virtual machines.

(2) *Download Hadoop*: Use the stable version 2.4.1 of Hadoop. The download link is https://hadoop.apache.org/.

(3) *Download JDK*: Use JDK version 1.7.0_45 (64-bit). Download the JDK package jdk-7u45-linux-x64.tar.gz from the Oracle official website. The download link is http://www.oracle.com/technetwork/java/javase/downloads/java-archive-downloads-javase7-521261.html.

(4) *Create a New User "hadoop"*: On each node, use the useradd command to create a new user named "Hadoop" and set a password for it.

```
useradd hadoop
passwd hadoop
```

(5) *Permanently Disable the Firewall on Each Node* (*Root Privileges Required*): Execute the following commands on each node to permanently disable the firewall:

```
chkconfig iptables off // This ensures the change is
    permanent and will not revert after a reboot
```

(6) *Configure SSH for Passwordless Access Between Hadoop Nodes*: SSH (Secure Shell) is a security protocol built on the application and transport layers, designed to provide security for remote login sessions and other network services. By using the SSH protocol, information leakage during remote management can be effectively prevented. The working principle of the SSH protocol is illustrated in Figure 8.16.

8.5.2. JDK installation and configuration

(1) Create a directory /usr/java on any one of the four virtual machines, and then extract the downloaded JDK package into this directory. The specific command line is as follows, where the z parameter calls the gzip compression program function, v displays the detailed decompression process, x extracts the file, and the f parameter is followed by the name of the file to be extracted:

```
mkdir /usr/java
tar -zxvf jdk-7u45-linux-x64.tar.gz -C /usr/java
```

server A	3. Server A sends a connection request to server B, including username, IP address, etc.	server B
1. First, generate a key pair on server A. 2. Copy the public key to server B and rename it to authorized_keys.		4. After receiving the information from server A, server B checks the authorized_keys for the corresponding username and IP address. If found, server B randomly generates a string: asdf.
	5. Server B encrypts the string asdf using the public key and sends it to server A.	
6. Upon receiving the message from server B, server A decrypts it using the private key and sends the decrypted string back to server B.		7. Server B compares the received decrypted string with the initially generated string; if they match, it grants passwordless access via authorized_keys.
Private Key Public Key		Public Key

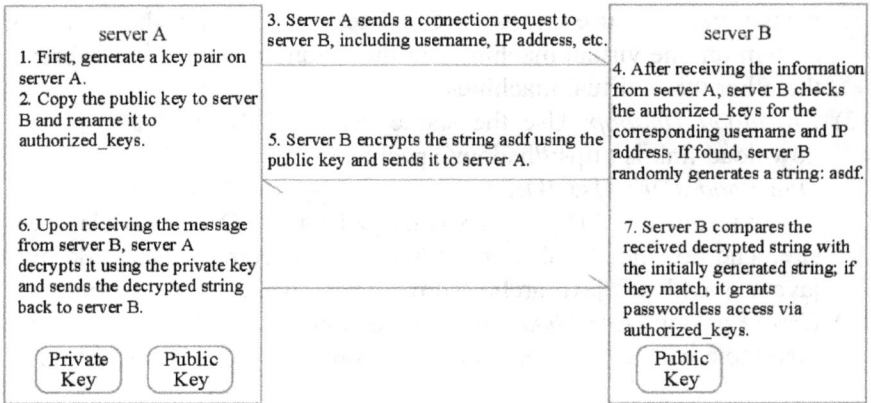

Fig. 8.16.   Working Principle of SSH.

(2) Configure the Java environment variables with the following commands:

```
#set java environment
export JAVA_HOME=/usr/java/jdk1.7.0_45
export JRE_HOME=/usr/java/jdk1.7.0_45/jre
export CLASSPATH=.:$JAVA_HOME/lib:$JRE_HOME/
    lib:$CLASSPATH
export PATH=$JAVA_HOME/bin:$JRE_HOME/bin:$PATH
```

(3) After saving the Java environment variables, enter the following command in the command line to make the environment variables take effect immediately.

```
source /etc/profile    // Make  the  environment
                          variable settings take effect
```

(4) Test if the JDK installation is successful using the which command:

```
which java
```

The system should display the following:

```
/usr/java/jdk1.7.0_45/bin/java
```

This indicates that the JDK configuration was successful.

8.5.3. *Installing Hadoop and configuring Hadoop environment variables*

The installation and configuration of Hadoop on each node in the Hadoop cluster are identical. You can perform the installation and configuration on one virtual machine and then copy the setup to the corresponding directories on the other nodes.

Place hadoop-2.4.1.tar.gz in the /home/hadoop directory on vm1 and then decompress it using the following commands:

```
cd /home/hadoop
tar -zxvf hadoop-2.4.1.tar.gz
```

Add the Hadoop installation path to /etc/profile. Append the following lines to the end of the file. This step needs to be performed on each node:

```
#set hadoop environment

export HADOOP_HOME=/home/hadoop/hadoop-2.4.1
export PATH=$PATH:$HADOOP_HOME/bin
```

After saving the Hadoop environment variables, execute the following command to apply the changes:

```
source /etc/profile    //  Make   the   environment
                           variable settings take effect
```

8.5.4. *Modifying Hadoop configuration files*

The Hadoop configuration files are located in the conf folder. You need to modify the following files within this folder: hadoop-env.sh, core-site. xml, hdfs-site.xml, mapred-site.xml, masters, and slaves.

(1) Modify hadoop-env.sh: The Java environment variables for Hadoop are set in hadoop-env.sh. Use vim to open the hadoop-env.sh file, find the location where the Java environment variable is set, and update it to the JDK installation path. Save and exit the file. The command is as follows:

```
export JAVA_HOME=/usr/java/jdk1.7.0_45
```

(2) Modify core-site.xml: The core-site.xml file is used to set the HDFS address and port number for the Hadoop cluster, as well as the path for the temporary files used by HDFS. When reformatting HDFS, files in the temporary folder should be deleted first.

Open core-site.xml using vim, and add the following configuration between the <configuration> and </configuration> tags:

```
<property>
<name>hadoop.tmp.dir</name>
<value>/home/hadoop/hadoop-2.4.1/tmp</value>
</property>
<property>
<name>fs.defaultFS</name>
<value>hdfs://192.168.122.101/:9000</value>
</property>
```

The IP address should be set to the IP address of the cluster's NameNode (Master) node, which in this case is 192.168.122.101.

(3) *Modify hdfs-site.xml*: This file specifies the number of replicas for HDFS data. Use the following configuration:

```
<property>
<name>dfs.replication</name>
<value>1</value>
</property>
```

(4) *Modify mapred-site.xml*: Configure the hostname and port for the JobTracker. The following settings should be used:

```
<property>
<name>mapreduce.jobtracker.address</name>
<value>http://192.168.122.101:9001</value>
<description>NameNode</description>
</property>
```

(5) *Modify masters file*: Open the masters file with vim and write the hostname of the NameNode (Master) node. In this case, it is vm1. Save and exit the file.

```
vm1
```

(6) *Modify slaves file*: Open the slaves file with vim and write the host-
names of the DataNode (Slave) nodes, which are vm2, vm3, and vm4.
Save and exit the file.

```
vm2
vm3
vm4
```

8.5.5. *Copying configured Hadoop files to other nodes and formatting*

At this point, Hadoop has been installed and configured on one node.
Since the installation and configuration are identical across all nodes in
the Hadoop cluster, you need to execute the following commands to copy
the Hadoop folder from vm1 to the other nodes:

```
scp -r /home/hadoop/hadoop-2.4.1 hadoop@vm2:/home/
    hadoop/
scp -r /home/hadoop/hadoop-2.4.1 hadoop@vm3:/home/
    hadoop/
scp -r /home/hadoop/hadoop-2.4.1 hadoop@vm4:/home/
    hadoop/
```

Before officially starting Hadoop, you need to format the HDFS with
the following command:

```
hadoop namenode -format
```

After successfully executing this formatting command, the following
message will be displayed:

```
18/03/28 21:21:20 INFO common.Storage: Storage
    directory /home/hadoop/ hadoop-2.4.1/tmp/dfs/name
    has been successfully formatted.
```

8.5.6. *Starting and stopping Hadoop*

Navigate to the /home/hadoop/hadoop-2.4.1/sbin/ directory to find several startup scripts:

```
distribute-exclude.sh mr-jobhistory-daemon.sh start-
    dfs.cmd stop-all.sh stop-yarn.sh

hadoop-daemon.sh   refresh-namenodes.sh   start-dfs.sh
    stop-balancer.sh yarn-daemon.sh
hadoop-daemons.sh slaves.sh start-secure-dns.sh stop-
    dfs.cmd yarn-daemons.sh
hdfs-config.cmd   start-all.cmd   start-yarn.cmd   stop-
    dfs.sh
hdfs-config.sh    start-all.sh    start-yarn.sh    stop-
    secure-dns.sh
httpfs.sh start-balancer.sh stop-all.cmd stop-yarn.cmd
```

To start Hadoop, execute the start-all.sh script:

```
cd /home/hadoop/hadoop-2.4.1/sbin/

./start-all.sh
```

To verify the startup of processes on the NameNode (192.168.122.101), use the jps command:

```
[root@host name local ~ ] # jps

11850 SecondaryNameNode
11650 NameNode
11949 JobTracker
12132 Jps
```

On the DataNodes (192.168.122.102, 192.168.122.103, 192.168.122.104), use the jps command:

```
[root@host name local ~ ] # jps
8727 DataNode
8819 TaskTracker
8958 Jps
```

At this point, Hadoop is successfully configured (note that configuration methods may vary for different Hadoop versions).

To stop Hadoop, use the following command:

```
cd /home/hadoop/hadoop-2.4.1/sbin/
./stop-all.sh
```

8.5.7. *Running the WordCount test program*

(1) Create a folder named WordCount in the current directory of the hadoop user, and then create two test files, file1.txt and file2.txt, within this folder. Populate these files with the following content:
file1.txt:

```
This is the first hadoop test program!
```

file2.txt:

```
This program is not very difficult, but this program
    is a common hadoop program!
```

(2) Create a directory named input in the HDFS and check its contents. Use the following commands:

```
hadoop fs -mkdir /input
hadoop fs -ls /
```

(3) Upload the file1.txt and file2.txt files from the WordCount folder to the newly created input folder in HDFS. Use the following command:

```
hadoop fs -put /home/hadoop/WordCount/*.txt /input
```

(4) Run the Hadoop example program WordCount using the following command:

```
hadoop   jar   hadoop-mapreduce-examples-2.4.1.jar
    wordcount intput output
```

(5) Check the directory information and the results of the WordCount program.

To view the directory information of the output results, use the following:

```
hadoop fs -ls /output
```

To view the WordCount results, use the following:

```
hadoop fs -cat /output/part-r-00000
```

The output will display the count of each word as follows:

```
This            2
a               1
common          1
difficult,but   1
first           1
hadoop          2
is              3
not             1
program         2
program!        2
test            1
the             1
this            1
very            1
```

This output shows the number of occurrences for each word.

Exercises

(1) What is the default size of a data block (Block) in HDFS?
(2) Draw a diagram of the basic architecture of HDFS and provide a brief overview of its principles.
(3) Provide a brief overview of the MapReduce programming model.
(4) What are the characteristics of the columnar database HBase?
(5) Set up a Hadoop development environment and implement it.

Chapter 9

Storm: A Topology-Based Real-Time Stream Processing Framework

Batch processing and stream processing are two modes of big data processing. As discussed in the previous chapter, Hadoop is a standard batch processing framework. In batch processing, data must first be collected and stored in a storage location, such as a database, before it can be analyzed and processed. Generally, batch processing is advantageous for performing complex analyses on large datasets; however, due to the large data volumes and complex processing involved, the results typically exhibit a relatively high latency. Stream processing, on the other hand, is a fundamentally different mode, characterized by a simpler processing workflow and lower latency, making it more suitable for real-time computation. Storm is a real-time data processing framework and an extremely effective open-source tool for real-time computing. This chapter provides a systematic introduction to Storm.

9.1. Introduction to Storm

Storm is an open-source, real-time computation platform, initially developed by Nathan Marz, an engineer at the social media data analytics company Backtype. It was later acquired by Twitter (renamed X) and contributed to the Apache Software Foundation, where it has since been elevated to the status of an Apache top-level project. The logo of Storm is shown in Figure 9.1.

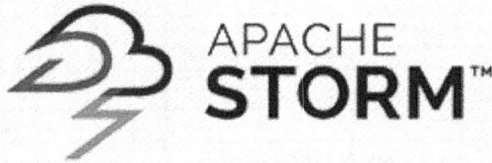

Fig. 9.1. Storm logo.

The implementation language for Storm is Clojure, a high-level, dynamic functional programming language based on the LISP programming language. Clojure includes a compiler that allows it to run on Java and .Net environments. Its implementation is very similar to LISP, a language renowned for its expressiveness and power, particularly in supporting functional programming styles.

Storm is a highly promising stream processing system that quickly gained traction in many major companies, including Taobao, Baidu, Twitter, and Groupon (a hybrid of e-commerce, Web 2.0, Internet advertising, and offline models). Storm simplifies the processing of unbounded stream data compared to traditional methods and is widely used in real-time analytics, online machine learning, continuous computation, and Distributed Remote Procedure Calls (DRPCs). For example, Ctrip's website performance monitoring system employs Storm. This system monitors the performance of Ctrip's website in real time using the Performance standards provided by HTML5 to obtain performance metrics, logs them, and analyzes the logs in real time via a Storm cluster, storing the results. The system uses DRPC to aggregate reports and triggers alert events based on historical data comparisons and predefined rules. Baidu, as one of the largest search engines, extensively uses Storm to process search logs and perform real-time analysis. The following introduces the basic concepts and features of Storm:

(1) *Core Concepts of Storm*: There are several essential core concepts (components) in Storm that need to be understood first, including Topology, Nimbus, Supervisor, Worker, Executor, Task, Spout, Bolt, Tuple, Stream, and Stream Grouping, as shown in Table 9.1.

Table 9.1.   Core concepts (components) of storm.

Component	Concept
Topology	The logical encapsulation of a real-time computation application within a Topology object, similar to a job in Hadoop. Unlike jobs, a Topology runs continuously until the process is terminated.
Nimbus	Responsible for resource allocation and task scheduling, similar to the JobTracker in Hadoop.
Supervisor	Receives tasks assigned by Nimbus, manages the start and stop of Worker processes, similar to the TaskTracker in Hadoop.
Worker	The component responsible for executing the actual logic.
Executor	As of Storm 0.8, an Executor is a physical process within a Worker. Multiple tasks belonging to the same Spout/Bolt can share a single Executor, but an Executor can only run tasks from the same Spout/Bolt.
Task	The specific work that each Spout/Bolt performs; it also serves as the unit for grouping between different nodes.
Spout	The component in a Topology that generates the data source. Typically, a Spout retrieves data from a data source, then calls the nextTuple function to send data for Bolts to consume.
Bolt	The component in a Topology that receives data from a Spout and processes it. Bolts can perform operations like filtering, function application, aggregation, and writing to databases. When a Bolt receives a message, it calls the execute function, where the user can implement the desired operations.
Tuple	The basic unit of message transmission.
Stream	A continuous flow of Tuples, which constitutes a Stream, also known as a data stream.
Stream Grouping	The method for grouping messages. Storm provides several useful grouping strategies, including Shuffle, Fields, All, Global, None, Direct, and Local or Shuffle.

(2) *Storm Data Streams*: As mentioned above, the data processed by Storm are referred to as a data stream (Stream). Within Storm, data streams are transmitted between components in the form of a series of Tuple sequences. The transmission process is illustrated in Figure 9.2. Each Tuple can contain different types of data, such as integers and

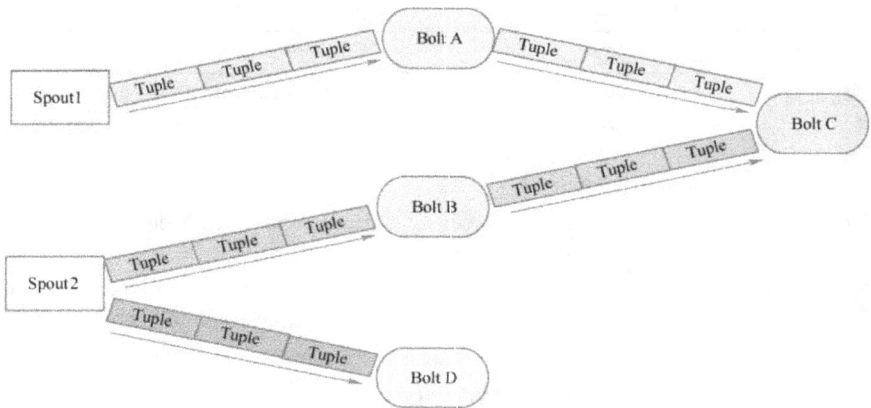

Fig. 9.2.   The process of data stream handling in storm.

strings, but the data types at corresponding positions between different Tuples must be consistent. This is because the data types within a
Tuple are predefined by each component before processing.

In a Storm cluster, each node can process hundreds or thousands
of Tuples per second. The data stream flows continuously from one
component to the next, much like water flowing through pipes, with
Tuples acting as the conduits carrying the data stream.

(3) *Reliability of Storm*: Storm ensures that each Tuple is fully processed
by the Topology. It tracks the message tree generated from each Tuple
sent by a Spout (a message tree is formed as Bolts process the
received messages and generate zero or more new messages, which
continues recursively). Storm guarantees that this message tree is successfully executed. Each Tuple is assigned a timeout; if Storm does
not detect the successful execution of a Tuple within the specified
time, it assumes that the Tuple has failed and resends it. This mechanism ensures that every message unit is completely processed within
a designated timeframe.

(4) *Features of Storm*: Storm is an open-source, distributed real-time
computing system designed to process large volumes of data streams
in a simple and reliable manner. It supports horizontal scalability,
offers high fault tolerance, and ensures that every message is processed with high speed (in a small cluster, each node can handle

millions of messages per second). Storm is easy to deploy and maintain, and importantly, it allows for the development of Storm-based applications using any programming language. This flexibility makes Storm a popular choice for real-time stream processing in the current big data environment. The following are some key features of Storm:

- *Integrity*: Storm employs the Acker mechanism to ensure that data are not lost, and it uses transactional mechanisms to guarantee data accuracy.
- *Fault Tolerance*: Storm has adaptive fault tolerance capabilities. Since Storm's daemons (Nimbus and Supervisor) are stateless and can quickly recover, users can restart them as needed. When a Worker fails or a machine experiences a fault, Storm automatically reallocates new Workers to replace the failed ones without causing additional impact.
- *Scalability*: Due to the parallel nature of Topologies, operations can be executed across machines or clusters. Components within a Topology can be configured with flexible parallelism settings, ensuring high throughput and low latency in Storm's data processing.
- *Ease of Use*: Storm requires minimal installation and configuration to deploy and start, and development is swift and user-friendly, making it accessible for users.
- *Free and Open Source*: Storm is a free and open-source project, allowing users to utilize it without payment. However, it is governed by the Eclipse Public License (EPL), which is a relatively permissive open-source license allowing users to open-source or keep proprietary their Storm applications.
- *Support for Multiple Languages*: While Storm is developed in Clojure and primarily interfaces through Java, it supports multiple programming languages. Storm provides adapters for various languages, including Ruby, Python, PHP, and Perl.

(5) *Application Scenarios of Storm*: Storm is used for real-time processing of massive amounts of continuously generated data, similar to a production assembly line. Its primary applications include the following:

- *Log Analysis*: Analyzing specific data from vast volumes of logs and storing the results in external storage to provide analytical support for business decision-making.

- *Pipeline Systems*: Transferring data from one system to another.
- *Message Transformation*: Converting received messages into a specific format and storing them in another system, such as a message middleware.

9.2. Storm Principles and Architecture

9.2.1. *Principles of the storm programming model*

A model is an abstraction of the common aspects of a phenomenon, and a programming model is an abstraction of the common aspects of programming. The most important common aspects in programming are the abstraction, organization, and reuse of code. The programming model does not consider the smallest operational units because some languages can operate at the bit level, which is at the same abstraction level as machine instructions. Additionally, programming models exist at a methodological or conceptual level and are often referred to as programming methods, programming approaches, programming patterns, or programming techniques.

Storm employs a programming model similar to the pipeline processing approach commonly used in daily tasks. The data stream (Stream) in Storm is an abstraction of data, represented as an unbounded sequence of Tuples over time. In a Topology, a Spout is the source of a Stream, responsible for emitting Streams from specific data sources into the Topology. Bolts can receive multiple Streams as input, process the data, and, if needed, emit new Streams for further processing by downstream Bolts. Figure 9.3 illustrates the complete topology for executing a task in Storm.

In a Topology, each computational component (Spout and Bolt) has a degree of parallelism that can be specified when creating the Topology. Storm allocates a corresponding number of threads within the cluster to

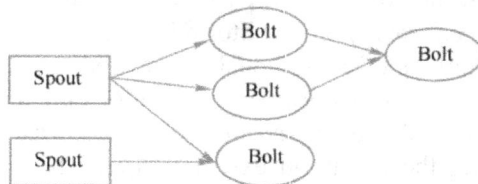

Fig. 9.3.　Topology of the storm programming model.

execute this component concurrently. This raises a question: Since both Spouts and Bolts have multiple Task threads running, how are Tuples sent between the two? Storm provides several data stream grouping (Stream Grouping) methods to address this issue. When defining a Topology, it is necessary to specify the type of Stream that each Bolt will receive as input (Spouts only send Streams and do not receive them). Storm offers the following seven data stream grouping methods:

(1) *Shuffle Grouping*: Data are distributed randomly among Tasks. This ensures that each Task on the same level of a Bolt processes a consistent number of Tuples, as shown in Figure 9.4.
(2) *Fields Grouping*: Tuples are grouped based on the value of one or more Fields within the Tuple. For example, Streams can be grouped by the value of user-id, so Tuples with the same user-id value are assigned to the same Task, as illustrated in Figure 9.5.
(3) *All Grouping*: Every Tuple is sent to all Tasks, as shown in Figure 9.6.
(4) *Global Grouping*: The Stream selects a single Task as the destination, typically the Task with the latest ID, as depicted in Figure 9.7.

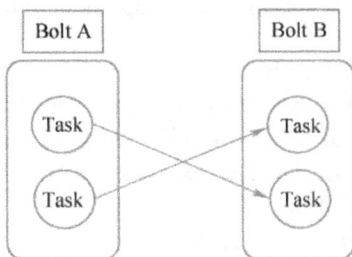

Fig. 9.4. Shuffle grouping random distribution pattern.

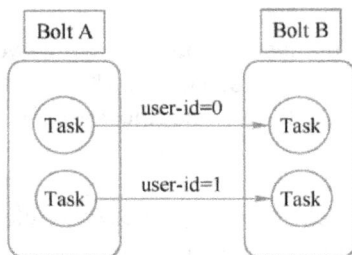

Fig. 9.5. Fields grouping pattern.

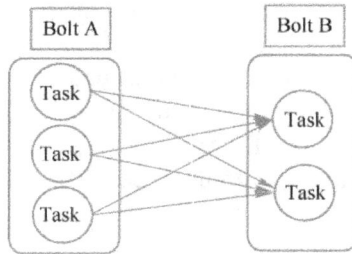

Fig. 9.6.   All grouping broadcast mode.

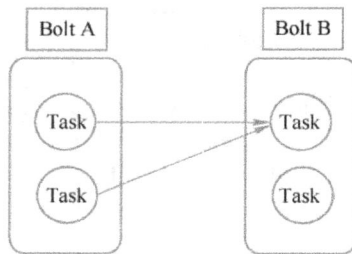

Fig. 9.7.   Global grouping single destination mode.

(5) *None Grouping*: Currently, this is equivalent to Shuffle Grouping.
(6) *Direct Grouping*: The Spout or Bolt that generates the data can deter-mine which Task of the Bolt will consume the Tuple. When using Direct Grouping, the emit Direct method of the OutputCollector is used to distribute the data.
(7) *Local or Shuffle Grouping*: If one or more Tasks of the target Bolt are in the same Worker process as the current Task generating the data, the Tuple is sent directly to the target Task within the current Worker process via internal thread communication.

For example, the process of performing a word count task using Storm is illustrated in Figure 9.8. The WordCount Topology consists of a Spout followed by three Bolts, and the specific execution process is described as follows:

(1) The Sentence Spout class emits a series of single-valued tuples, named "sentence", containing a string value. An example is the following code:

```
{"sentence":my cat has fleas"}
```

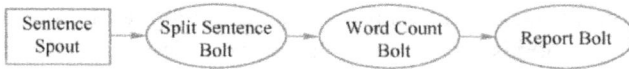

Fig. 9.8. WordCount topology.

For simplicity, the data source here is a static list of sentences. The Sentence Spout iterates through these sentences, emitting a tuple for each one. In a real application, a Spout typically connects to a dynamic data source, such as querying tweets from the Twitter API.

(2) Split Sentence Bolt subscribes to the tuple stream from the Sentence Spout. For each tuple it receives, it retrieves the value of the "sentence" object, splits it into individual words, and emits a tuple for each word.

(3) Word Count Bolt subscribes to the output of the Split Sentence Bolt and continually counts the occurrences of specific words. Each time it receives a tuple, it increments the counter associated with the word and emits the word along with its current count.

(4) Report Bolt subscribes to the output of the Word Count Bolt and maintains a table containing all words and their respective counts. Similar to the Word Count Bolt, when the Report Bolt receives a tuple, it updates the table and prints the contents to the console.

9.2.2. *Storm architecture*

Storm employs a Master/Slave architecture, where the master node is Nimbus and the slave nodes are Supervisors. The architecture is illustrated in Figure 9.9. In a traditional Master/Slave architecture, the Master node

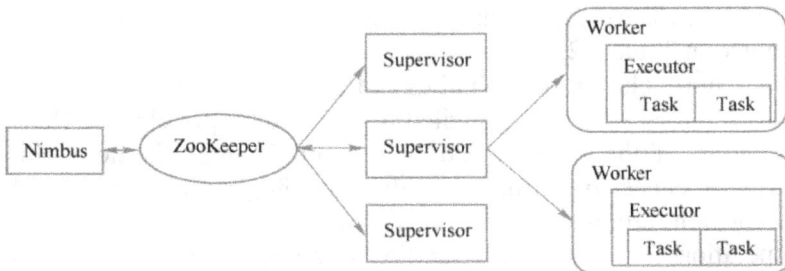

Fig. 9.9. Storm architecture.

is responsible for tasks such as receiving, assigning, and monitoring jobs, while the Slave nodes are responsible for executing the tasks. Generally, the Master/Slave architecture in Storm follows these principles.

The master node, Nimbus, is responsible for distributing the task (Topology) code across the cluster and for monitoring the system. Upon receiving a task, the slave node, Supervisor, will start one or more processes (referred to as Workers) to handle the task. In essence, the task is ultimately assigned to Workers.

Both Nimbus and Supervisor nodes can recover quickly from failures, and both are stateless. On each Supervisor node, multiple Workers can be launched, and within each Worker, multiple Executors can be initiated. Each Executor can then be further divided into multiple Tasks, with a Task being the smallest unit of work allowed by the system. Storm uses Zookeeper to store metadata for Nimbus, Supervisors, and the internal Workers within Storm, allowing for fault recovery.

At first glance, the Storm cluster is very similar to the Hadoop cluster. The key difference is that Hadoop runs MapReduce jobs, while Storm runs Topologies (an abstraction of tasks and a core component of Storm). Another significant difference is that Hadoop's MapReduce jobs eventually terminate, whereas Storm's Topologies are designed to run continuously.

Comparing Storm with Hadoop, Nimbus's role is similar to Hadoop's JobTracker. Nimbus is responsible for distributing the code across the cluster, assigning work to other nodes, and monitoring the status of these nodes. Each worker node runs a Supervisor process, which listens for tasks assigned by Nimbus and starts or stops specific Worker processes as needed. Each Worker process executes a specific Topology, and within the Worker process, execution threads are called Executors. Each Executor can contain one or more Tasks, with Task being the smallest processing unit in Storm. The comparison between Storm and Hadoop components is illustrated in Table 9.2.

In a Storm cluster, both Nimbus and Supervisor are stateless, with all coordination between them handled by the ZooKeeper cluster. The state information of the Nimbus and Supervisor daemons is stored in the ZooKeeper cluster or on the local disks of the respective nodes where these daemons are running. This architecture ensures high stability during the operation of the Storm cluster. For instance, if the command "kill -9 <pid of num>" is used to terminate a Nimbus or Supervisor process, ZooKeeper will immediately initiate a backup Nimbus or Supervisor,

Table 9.2. Comparison between Storm and Hadoop components.

	Storm	**Hadoop**
System roles	Nimbus	JobTracker
	Supervisor	TaskTracker
	Worker	Child
Application names	Topology	Job
Component interfaces	Spout/Bolt	Mapper/Reducer

allowing the Storm cluster to maintain its current state and continue functioning after a restart.

9.3. Storm-Yarn Introduction

Storm-Yarn is built on top of Storm, with a framework structure and data processing method that are largely consistent with Storm. The key difference is that Storm-Yarn integrates the components of Storm with the various functional parts of Hadoop's resource manager, Yarn.

9.3.1. *Background of storm-yarn*

Storm is highly capable of real-time data processing, leveraging a distributed architecture inspired by Hadoop to overcome the bottleneck of single-node processing limitations. By adopting Hadoop's master-slave architecture for managing distributed clusters, Storm inherently possesses scalability and fault tolerance.

Hadoop, having been around longer, has developed a mature and refined underlying system with its HDFS distributed file system and MapReduce distributed processing framework. Supported by a variety of upper-layer applications, Hadoop has established itself as a core technology in the field of distributed processing and a *de facto* standard in big data technology. Its distributed ecosystem has received significant backing from major internet companies like Google, Alibaba, and Cloudera. The redefined Yarn framework in Hadoop 2.x provides automated resource management for upper-layer applications, greatly simplifying the management of distributed application resources.

Storm-Yarn integrates Storm's real-time processing capabilities into the Hadoop ecosystem, enabling Storm to access Hadoop's storage resources (such as HDFS, HBase, and Hive) and fully utilize cluster computing resources for more extensive real-time data processing.

Advantages of Storm-Yarn based on Hadoop:

(1) *System Elasticity*: Storm's real-time processing characteristic often leads to variable processing loads due to the differing characteristics and volumes of data streams, making it difficult to predict load specifics accurately. Deploying Storm clusters on Hadoop's Yarn framework allows for elastic scaling of system resources, automatically acquiring and releasing unused resources on Hadoop as needed, thereby enhancing overall cluster resource utilization.

(2) *Data Sharing and Application Migration*: Storm-Yarn meets the need for big data processing that involves both real-time and batch processing of the same dataset. For example, real-time data generated by users can be processed immediately using Storm's low-latency capabilities. If the same data require further analysis later, they can be stored temporarily and processed offline using MapReduce to extract valuable insights. This approach allows for multifaceted utilization of the same data.

9.3.2. *Storm-yarn architecture*

The architecture of Storm-Yarn maintains the core functionality of components from Storm but introduces a clear separation of these components to integrate effectively with Yarn. The architecture of Storm-Yarn is illustrated in Figure 9.10.

When the Storm Master application initializes, it launches two services within the same container: the Storm Nimbus Server and the Storm UI Server. It then requests resources from the Yarn Resource Manager based on the number of Supervisors needed. In the current implementation, the Storm Master requests all available resources on a node to start the Supervisor service. This means that each Supervisor will exclusively occupy the node without sharing its resources with other services, thereby minimizing interference from other services on the Storm cluster.

The operation flow of Storm-Yarn is as follows:

(1) Storm-Yarn first requests the Yarn Resource Manager to start a Storm Master application, as shown in step ① of Figure 9.10.

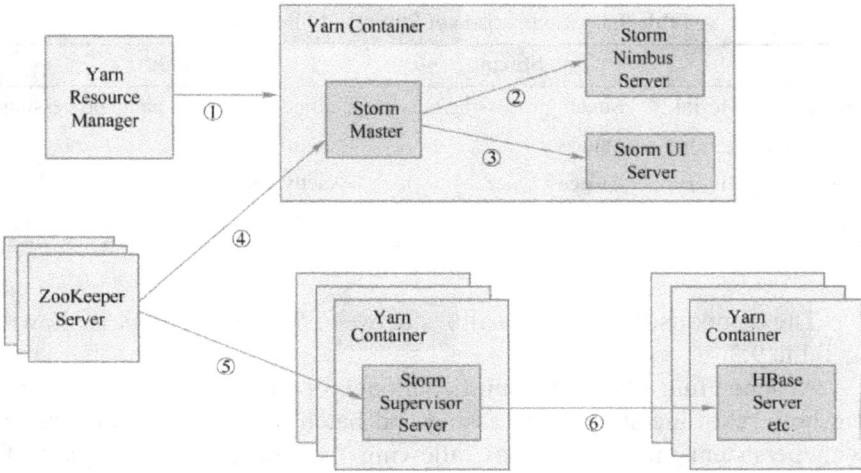

Fig. 9.10. Storm-Yarn architecture.

(2) The Storm Master then launches the Storm Nimbus Server and Storm UI Server locally, as indicated in steps ② and ③ of Figure 9.10.

(3) Zookeeper Server is used to maintain the master-slave relationship between Nimbus and Supervisor within the Storm-Yarn cluster, as depicted in steps ④ and ⑤ of Figure 9.10. Nimbus and Supervisor run in separate resource containers (Yarn Containers) allocated by the Yarn Resource Manager.

Additionally, Storm-Yarn can also operate or access distributed databases such as HBase running on Hadoop, as shown in step ⑥ of Figure 9.10.

For more details on the latest developments in Storm-Yarn, visit https://github.com/yahoo/storm-yarn.

9.4. Comparison between Storm and Flink

Apache Flink is a framework and distributed processing engine designed for stateful computations on unbounded and bounded data streams. Flink can operate in various cluster environments and perform computations at in-memory speeds and any scale. In contrast to the Storm framework, Flink innovatively introduces a unified stream and batch processing model.

Table 9.3.   Comparison of Storm and Flink features.

	Storm	Flink
Processing Model	Stream processing only	Unified stream and batch processing
State Management	Stateless	Stateful
Message Delivery	At Least Once	Exactly Once
Fault Tolerance	ACK mechanism	Checkpoint mechanism

The comparison of functionalities between Storm and Flink is shown in Table 9.3.

Apache Flink adheres to a unified model for data processing, supporting both real-time stream processing and batch processing. It integrates with persistent message queues, allowing for near-unlimited replay of data streams (such as Apache Kafka or Amazon Kinesis). Flink treats batch processing as operations on bounded datasets and real-time stream data as unbounded datasets. Users can process both real-time streaming data and static historical datasets using a single system. Flink is a unified framework for both stream and batch processing. Due to its pipeline-based data transmission between parallel tasks, Flink supports both streaming and batch processing at runtime. Once data are transmitted to the processing queue, batch processing tasks can be employed to handle these blocked data.

In contrast, Storm exclusively processes streaming data and lacks batch processing capabilities. Although Flink's streaming processing engine is similar to Storm's, such as Flink's parallel tasks and Storm's bolts, which both aim to reduce data latency through pipeline data transfer, Flink offers more advanced APIs compared to Storm. For instance, Flink's DataStream API provides functionalities like Map, GroupBy, Window, and Join, which replace Storm's bolts, reducing the need for custom implementation by programmers in Storm.

Regarding message delivery and fault tolerance, Storm provides "at-least-once" semantics, while Flink offers "exactly-once" guarantees. Storm employs the ACK mechanism, whereas Flink uses a checkpointing method. Specifically, when data sources inject markers periodically and feed them into the data stream, each task executor checks its internal state upon receiving a marker. If all data outputs for a marker are received, it is confirmed that the marker has been completely processed. If any segment of data is missing, all source operators reset their processing to the most recently confirmed marker and continue execution. This checkpointing approach is lighter compared to ACK. However, Storm has also introduced "exactly-once" functionality in later developments, based on micro-batch processing.

Since its introduction, Flink's advanced concepts have quickly garnered attention from major Internet companies and the open-source community. As a next-generation big data stream processing framework, Flink provides millisecond-level latency while ensuring, high throughput, and accuracy of results. It also offers rich support for time types, window computations, exactly-once semantics, state management, and Complex Event Processing (CEP). Flink's advantages in real-time analysis have prompted more companies to migrate their real-time projects to Flink, with its community rapidly growing. Flink has become a significant platform for real-time applications across major companies. Further learning and understanding of Flink based on Storm is highly beneficial.

9.5. Setting Up the Storm Development Environment

Storm offers two operational modes: local mode and remote mode. In local mode, you can develop and test Topologies entirely within processes on your local machine. In remote mode, you need to submit Topologies to a cluster. Setting up a Storm development environment involves installing the necessary dependencies for Storm, enabling you to develop and test Storm Topologies in local mode, package the Topologies, and deploy them to a remote Storm cluster. Additionally, you need to install the Storm system toolkit. This section covers setting up the Storm development environment on a cluster consisting of four nodes.

9.5.1. *Storm installation instructions*

(1) *Environment Specifications*:

- *Operating System*: CentOS 64-bit.
- *Cluster Configuration*: four nodes with IP addresses ranging from 192.168.10.100 to 192.168.10.103 or configured according to the actual cluster setup.

(2) *Software Installation Instructions*:

- *Required Software*: JDK, Python, gcc–c++, uuid*, libuuid, lib-tool, and ibuuid-devel.
- *Storm Required Toolkits*: ZooKeeper, ZeroMQ, JZMQ, and Storm.
  - *ZooKeeper*: ZooKeeper is a distributed, open-source coordination service for distributed applications. It is an open-source implementation of Google's Chubby and a critical component for Hadoop and HBase. ZooKeeper provides consistency services for distributed applications, including configuration maintenance, domain name service, distributed synchronization, and group services. Its purpose is to encapsulate complex and error-prone services, offering users simple interfaces with high performance and reliable functionality. Download ZooKeeper from https://zookeeper.apache.org/.
  - *ZeroMQ*: ZeroMQ is a high-performance asynchronous messaging library designed for scalable distributed or concurrent applications. It provides a messaging queue but does not require a dedicated message broker. The library is designed with a socket-style API, offering various socket types that represent many-to-many connections between ports. It operates at the message level with optimized message patterns. Download ZeroMQ from http://zeromq.org/.
  - *JZMQ*: JZMQ is the Java version of ZeroMQ, implemented through JNI for maximum performance, specifically as the Java binding for ZeroMQ. Download JZMQ from http://zeromq.org//bindings:java/.
  - *Storm*: The main program for the Storm system. This section uses Storm version 0.8.1. Download Storm from http://storm.apache.org/.

9.5.2. *Storm installation*

First, install Storm on one of the nodes in the cluster. Before installing Storm, it is assumed that the user has already configured the CentOS operating system and established the root superuser. This section simplifies the installation and configuration steps for CentOS.

Note: The following installation steps are performed under the root user, and the required tool packages are stored in the compressed file "storm.tar.gz".

(1) *Preparation for Installation*: Extract the "storm.tar.gz" compressed file using the following command:

```
tar -zxvf storm.tar.gz
```

Change the current working directory to the extracted Storm directory using the following command:

```
cd storm
```

(2) *Installing Dependencies and Software*: Use the yum command to install the dependency packages g++, uuid*, libtool, libuuid, and libuuid-devel as follows.

Use the yum command to install the required dependencies: gcc-c++, uuid*, libtool, libuuid, and libuuid-devel, as shown in the following:

```
yum -y install gcc-c++、uuid*、libtool、libuuid、
    libuuid-devel
```

Note: The dependency installation command uses yum. Different Linux distributions use different installation commands; for example, Ubuntu uses apt-get. Users should use the appropriate command based on their chosen operating system. Some dependency tools may have different names on different Linux systems, but corresponding packages can always be found.

Install the JDK (here using the rpm method) with the following commands:

```
chmod 755 jdk-7u71-linux-x64.rpm
rpm -ivh jdk-7u71-linux-x64.rpm
```

After the JDK installation is successful, it will be stored by default in the "/usr/java" directory.

Use the "vim" or "vi" editor to open the "/etc/profile" file and configure the environment variables with the following commands:

```
#set java environment
JAVA_HOME=/usr/java/jdk1.7.0_71
```

```
JRE_HOME=/usr/java/jdk1.7.0_71/jre
PATH=$PATH:$JAVA_HOME/bin:$JRE_HOME/bin
CLASSPATH=.:$JAVA_HOME/lib/dt.jar:$JAVA_HOME/lib/
   tools.jar:$JRE_HOME/lib
export JAVA_HOME JRE_HOME PATH CLASSPATH
```

Save and exit the editor, then use the following command to immediately apply the environment variables:

```
source /etc/profile
```

Check if the installation was successful by running the command "java –version". As shown in Figure 9.11, this will display the system information after the successful JDK installation.

(3) *Install ZooKeeper*: Move the ZooKeeper installation package to the system directory using the following command:

```
cp –R zookeeper–3.4.5 /usr/local
```

You can add a symbolic link to this folder with the following command:

```
ln –s /usr/local/zookeeper3.4.5/ /usr/local/zookeeper
```

Open the "etc/profile" file using "vim" or "vi" to modify the configuration file with the following commands:

```
export ZOOKEEPER_HOME="/path/to/zookeeper"
export PATH=$PATH:$ZOOKEEPER_HOME/bin
```

Finally, create two directories for ZooKeeper to store temporary files and log files during operation, using the following commands:

```
mkdir /tmp/zookeeper
mkdir /var/log/zookeeper
```

ZooKeeper installation is now complete.

```
[root@storm ~]# java -version
java version "1.7.0_71"
Java(TM) SE Runtime Environment (build 1.7.0_71-b14)
Java HotSpot(TM) 64-Bit Server VM (build 24.71-b01, mixed mode)
[root@storm ~]#
```

Fig. 9.11. System information after successful JDK installation.

(4) *Install ZeroMQ*: Navigate to the software package directory with the following command:

```
cd zeromq-2.1.7
```

Configure the environment and install the ZeroMQ software package with the following commands:

```
./conf?igure
make
make install
```

During "./configure", the system will check if "JAVA_HOME" is correctly set. If not, an error will occur.

Update the dynamic link library using the following command:

```
Ldconfig
```

ZeroMQ installation is now complete.

(5) *Install JZMQ*: Navigate to the software package directory with the following command:

```
cd jzmq
```

Configure the environment and install the JZMQ software package with the following commands:

```
./autogen.sh
./conf?igure
make
make install
```

JZMQ installation is now complete.

(6) *Install Storm*: Extract the Storm compressed package using the following command:

```
unzip storm-0.8.1.zip
```

If the "unzip" command is not available, install it using the "yum" command as shown in the following:

```
yum -y install unzip
```

Move the extracted directory to the system installation directory using the following command:

```
mv storm-0.8.1 /usr/local
```

You can add a symbolic link to this directory with the following command:

```
ln -s /usr/local/storm-0.8.1 /usr/local/storm
```

Open the "etc/profile" file using "vim" or "vi" to configure the Storm environment variables with the following commands:

```
#set storm environment
export STORM_HOME=/usr/local/storm-0.8.1
export PATH=$PATH:$STORM_HOME/bin
```

Save and exit, then execute the following command to apply the environment variables immediately:

```
source /etc/profile
```

At this point, the installation of Storm-related software and tools on one node is complete.

Repeat the above steps on the remaining three nodes to complete the installation of Storm on each node.

9.5.3. *Storm configuration*

After installing Storm, follow these steps to configure it:

(1) *Configure Zookeeper*: Edit the ZooKeeper configuration file. Note that this step is not necessary for a single-node setup.

```
vim /usr/local/zookeeper/conf/zoo.cfg
```

Add the following lines at the end of the configuration file:

```
server.1=192.168.10.100:2888:3888
server.2=192.168.10.101:2888:3888
```

Save and exit. ZooKeeper configuration is now complete.

(2) *Configure Storm*: The purpose of this step is to configure the "storm. yaml" file since it does not include IP addresses by default. This

configuration allows communication between ZooKeeper nodes. Note that this must be done on all four nodes.

```
vim /usr/local/storm/conf/storm.yaml
```

Replace the following lines in the "storm.yaml" file,

```
# storm.zookeeper.servers:
# − "server1"
# − "server2"
```

with

```
storm.zookeeper.servers:
− "192.168.10.100"
− "192.168.10.101"
```

Replace

```
# nimbus.host: "nimbus"
```

with

```
nimbus.host: "192.168.10.100"
```

Add the directory for Storm's temporary files:

```
storm.local.dir: "/tmp/storm"
```

Additionally, configure the number of Workers that each Supervisor node can run. Each Worker occupies a separate port to receive messages. This option defines which ports can be used by Workers. By default, each node can run four Workers on ports 6700, 6701, 6702, and 6703. Adjust the number of ports based on node performance as needed.

```
supervisor.slots.ports:
− 6700
− 6701
− 6702
− 6703 (#Use the space key for line breaks, not the
  Tab key)
```

This completes the configuration for one node. For the remaining nodes, repeat the process. You can copy the configured "storm.yaml" file to the "/usr/local/storm/conf" directory on the other nodes, overwriting the existing "storm.yaml" file.

9.5.4. *Starting storm*

After configuring Storm, you can start the Storm processes with the following steps:

(1) *Start the Nimbus Process*: "bin/storm nimbus".
(2) *Start the Supervisor Process*: "bin/storm supervisor".
(3) *Start the UI Process* (*The UI process is a web-based graphical management interface for Storm. Once started, users can view Storm's system status via a web browser*): "bin/storm ui".
(4) *Start the Log Viewer Process*: "bin/storm logviewer".

On the Storm Nimbus node, the required processes are Nimbus, UI, and Log Viewer. On the Storm Supervisor nodes, the required processes are Supervisor and Log Viewer.

9.5.5. *Common storm command operations*

Storm provides a set of simple and user-friendly commands for managing topologies, allowing you to submit, destroy, deactivate, and rebalance topology tasks.

(1) *Submit Task Command*: "storm jar".
   Command Format: "storm jar [jar_path] [Topology_package_name.Topology_class_name] [Topology_name]".
   Example:

```
bin/storm jar examples/storm-starter/storm-
    starter-topologies-0.10.0.jar storm.starter.
    WordCountTopology wordcount
```

(2) *Destroy Task Command*: "storm kill".
   Command Format: "storm kill [Topology_name] -w [wait_seconds]" (*Note*: The -w flag specifies the waiting time after the topology is deactivated).
   Example:

```
storm kill topology-name -w 10
```

(3) *Deactivate Task Command*: "storm deactivate"
Command Format: "storm deactivate [Topology_name]".
Example:

```
storm deactivate wordcount
```

This command can suspend or deactivate a running topology. When deactivating a topology, all tuples that have been dispatched will be processed, but the "nextTuple" method of spouts will not be called. To destroy a topology, use the "kill" command, which safely destroys the topology by first deactivating it, allowing it to complete its current data stream during the waiting period.

(4) *Activate Task Command*: "storm activate"
Command Format: "storm activate [Topology_name]".
Example:

```
storm activate wordcount
```

(5) *Rebalance Task Command*: "storm rebalance"
Command Format: "storm rebalance [Topology_name]".
Example:

```
storm rebalance wordcount
```

This command redistributes the cluster tasks. It is commonly used when new nodes are added to a running cluster. The rebalance command will deactivate the topology, wait for the specified timeout, then redistribute the worker nodes, and restart the topology.

9.6. Storm Application Practice

This section demonstrates how to use Storm through practical examples. Storm provides a sample project called "storm-starter", which is a module designed for learning and utilizing Storm, containing various topology examples. This project can be obtained from "https://github.com/nathanmarz/storm-starter". Here, we introduce the "WordCountTopology" example from this sample project, which demonstrates how to use Storm to count the occurrences of each word in a file. Before delving into "WordCountTopology", we first introduce the use of Maven, a project management tool.

9.6.1. *Managing storm-starter with Maven*

(1) *Overview of Maven*: Maven is a project management and build automation tool provided by the Apache Software Foundation. It configures projects using "pom.xml" (Project Object Model), which details the various JAR files required to run the example code within the "storm-starter" directory. In addition to its program build capabilities, Maven offers advanced project management features. Due to Maven's reusable default build rules, simple projects can often be constructed with just a few lines of Maven build scripts.

Since submitting topologies in Storm must be performed on the master node, Maven needs to be installed on the master node.

(2) *Maven Installation*: Switch to the master node at "192.168.10.100", and use the "storm" user to navigate to the Storm directory with the following commands:

```
su – storm
cd storm
```

Download Maven using the following command:

```
wget http://mirrors.tuna.tsinghua.edu.cn/apache/
    maven/maven-3/3.5.0/binaries/apache-maven-3.5.0-
    bin. tar.gz
```

Extract the Maven package with the following command:

```
tar -zxvf apache-maven-3.5.0-bin.tar.gz
```

Move the extracted files to the system directory using the following command:

```
cp apache-maven-3.5.0-bin /usr/local
```

For convenience, rename the directory to a shorter name:

```
mv /usr/local/apache-maven-3.5.0-bin /usr/local/
    maven
```

Use "vim" or "vi" to open "/etc/profile" and configure the Maven environment variables by adding the following lines:

```
#maven
MAVEN_HOME=/usr/local/maven
```

```
PATH=$PATH:$MAVEN_HOME/bin
export MAVEN_HOME PATH
```

Save and exit, then apply the environment variables immediately with the following command:

```
source /etc/profile
```

To verify the successful installation of Maven, run the following:

```
mvn —version
```

This completes the Maven installation.

(3) *Managing the Example Project Storm-Starter with Maven*: Navigate to the "storm-starter" directory using the following command:

```
cd storm—starter
```

Run the compile command to package the project into a JAR file:

```
mvn —f m2—pom.xml package
```

After execution, a "storm-starter-0.8.1.jar" file will be generated in the "storm-starter/target" directory, as shown in Figure 9.12.

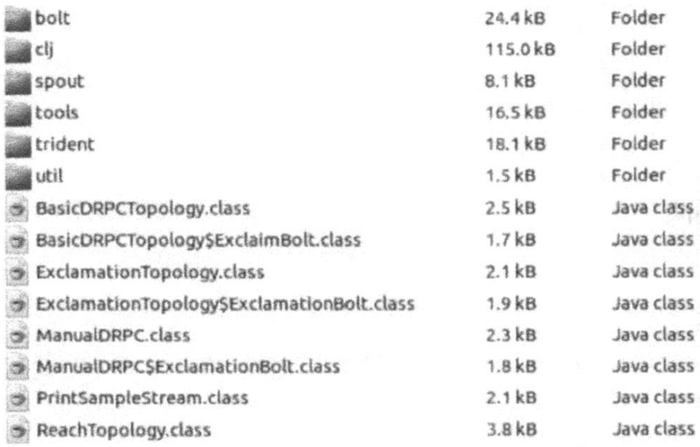

bolt	24.4 kB	Folder
clj	115.0 kB	Folder
spout	8.1 kB	Folder
tools	16.5 kB	Folder
trident	18.1 kB	Folder
util	1.5 kB	Folder
BasicDRPCTopology.class	2.5 kB	Java class
BasicDRPCTopology$ExclaimBolt.class	1.7 kB	Java class
ExclamationTopology.class	2.1 kB	Java class
ExclamationTopology$ExclamationBolt.class	1.9 kB	Java class
ManualDRPC.class	2.3 kB	Java class
ManualDRPC$ExclamationBolt.class	1.8 kB	Java class
PrintSampleStream.class	2.1 kB	Java class
ReachTopology.class	3.8 kB	Java class

Fig. 9.12. Directory of the storm—starter—0.8.1.jar file.

(4) *Submitting the Topology from Storm-Starter*: Run the Storm program, using the following format to submit a Topology:

```
storm jar all-my-code.jar backtype.storm.MyTopology
    arg
```

The file "all-my-code.jar" represents the name of the JAR package to be submitted, and "backtype.storm.MyTopology" denotes the topology name within the JAR package to be executed. The argument "arg" specifies the name to be used for the running topology; if left empty, it defaults to the topology;'s predefined name.

Navigate to the "/target" directory using the following command:

```
cd target
```

To submit the WordCount topology, use the following command:

```
storm jar storm-starter-0.0.1-SNAPSHOT-jar-with-
    dependencies.jar storm. starter.WordCountTopology
    wordcountTpy
```

The output will be as follows:

```
the       3
cow       1
jumped    1
over      1
moon      1
an        1
apple     1
a         1
day       1
keeps     1
doctor    1
away      1
four      1
score     1
and       2
seven     2
years     1
ago       1
```

snow	1
white	1
dwarfs	1
i	1
am	1
at	1
two	1
with	1
nature	1

(5) *Storm UI*: The program is compiled into a JAR file using Maven, and after submitting the application on the Nimbus node, relevant runtime information can be viewed on the Storm UI. The Storm UI includes four sections: Cluster Summary, Topology Summary, Supervisor Summary, and Nimbus Configuration. A screenshot of the Storm UI is shown in Figure 9.13.

   (1) *Cluster Summary*: Provides information about the entire cluster, including the total number of slots and their usage. By examining the number of free slots, users can estimate the Storm capacity and determine if the cluster needs to be scaled.

   (2) *Topology Summary*: Details the running topologies on the Storm cluster. When a user selects a specific topology, they can view all Spouts, Bolts, and associated statistics for that topology.

   (3) *Supervisor Summary*: Shows the status of Supervisor nodes across the Storm cluster. The "Uptime" indicates the duration since the Supervisor process was started.

Storm UI

Cluster Summary

Version	Nimbus uptime	Supervisors	Used slots	Free slots	Total slots	Executors	Tasks
0.8.1	2h 14m 57s	4	0	16	16	0	0

Topology summary

Name	Id	Status	Uptime	Num workers	Num executors	Num tasks

Supervisor summary

Host	Uptime	Slots	Used slots
node1	2h 14m 12s	4	0
node2	2h 14m 34s	4	0
node3	2h 14m 9s	4	0
node4	2h 13m 52s	4	0

Fig. 9.13.   Storm UI interface diagram.

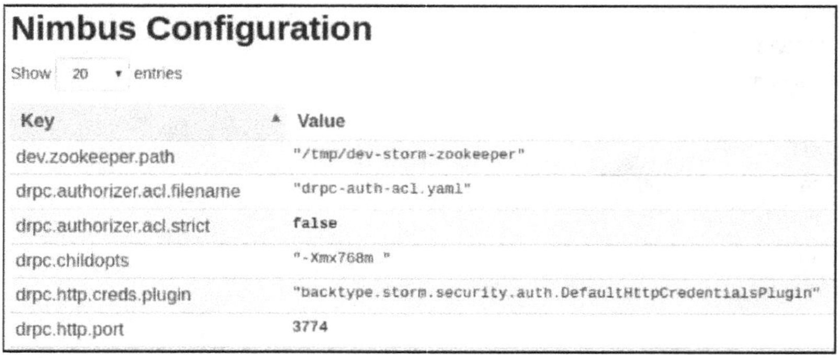

Nimbus Configuration

Show 20 ▼ entries

Key ▲	Value
dev.zookeeper.path	"/tmp/dev-storm-zookeeper"
drpc.authorizer.acl.filename	"drpc-auth-acl.yaml"
drpc.authorizer.acl.strict	false
drpc.childopts	"-Xmx768m "
drpc.http.creds.plugin	"backtype.storm.security.auth.DefaultHttpCredentialsPlugin"
drpc.http.port	3774

Fig. 9.14. Storm configuration information.

(4) *Nimbus Configuration*: Displays configuration information for the Storm cluster. If all nodes use the same configuration, this configuration effectively represents the settings for the entire Storm cluster, as shown in Figure 9.14.

9.6.2. *Analysis of WordCountTopology Source Code*

"WordCountTopology" is a typical learning example for understanding Storm. The main code segments of "WordCountTopology" are as follows:

```
// Create a TopologyBuilder object named builder
TopologyBuilder builder = new TopologyBuilder();
// Create a TopologyBuilder object named builder
builder.setSpout("spout", new RandomSentenceSpout(),
    5);
// Set up the Bolt component to split sentences,
    with 8 parallel instances
builder.setBolt("split", new SplitSentence(),
    8).shuffleGrouping("spout");
// Set up the Bolt to count word occurrences
builder.setBolt("count", new WordCount(), 12).
    fieldsGrouping("split", new Fields("word"));
```

This section provides a detailed explanation of this example to enhance understanding of Storm's execution flow and help in writing Topology programs.

Spout is the data source. A Spout is created using "TopologyBuilder" to simulate the data source:

```
builder.setSpout("spout", new RandomSentenceSpout(),
  5);
```

The business logic of the Spout is implemented in "RandomSentenceSpout", which includes an "open()" method to initialize the Spout. The specific code is as follows:

```
@Override
public void open(Map conf, TopologyContext context,
  SpoutOutputCollector collector) {
 _collector = collector;
 _rand = new Random();
}
```

When "nextTuple" is called, Storm requests the Spout to send a Tuple to the "OutputCollector". If no Tuple is available to send, this method returns to the "open()" method. The methods "nextTuple()", "ack()", and "fail()" must operate within a single thread in the Spout's task. If no Tuple is available, the method will sleep for a short period. The specific code is as follows:

```
@Override
public void nextTuple() {
Utils.sleep(100);        // Sleep for 100 milliseconds
 // Simulated data source
 String[] sentences = new String[]{ "the cow jumped
   over the moon", "an apple a day keeps the doctor
   away", "four score and seven years ago", "snow
   white and the seven dwarfs", "i am at two with
   nature" };
 // Randomly select a sentence
 String sentence = sentences[_rand.nextInt(sentences.
   length)];
 _collector.emit(new Values(sentence));
}
```

The declareOutputFields() method in Bolt is used to declare the fields contained in the current Tuple sent by Bolt:

```
@Override
public void declareOutputFields(OutputFieldsDeclarer
   declarer) {
   declarer.declare(new Fields("word"));
}
```

The "prepare()" method is similar to the "open()" method in Spout, providing the "OutputCollector" for the Bolt to send Tuples. This method is executed in the "execute()" method:

```
void prepare(Map stormConf, TopologyContext context,
   OutputCollector collector);
```

The "SplitSentence" class processes the Tuples sent by the Spout, splitting sentences into individual words. This processing is implemented using a Python script. The Java implementation is as follows:

```
public static class SplitSentence extends ShellBolt
   implements IRichBolt {

   public SplitSentence() {
     super("python", "splitsentence.py");
     // Use the parent class constructor, indicating
        that this method is implemented in Python, and
        the implementation file is named "splitsentence.
        py".
   }
   @Override
   public void declareOutputFields(OutputFields
     Declarer declarer) {
     declarer.declare(new Fields("word"));
    }
   // Used to declare special configuration settings
      for the current component; returns null
   @Override
   public Map<String, Object>
     getComponentConfiguration() {
       return null;
  }
}
```

The "splitsentence.py" Python script implementation is as follows:

```python
import storm
class SplitSentenceBolt(storm.BasicBolt):
    def process(self, tup):
        # Define a function to split the string into
        words
        words = tup.values[0].split(" ")      # Split
        the string by spaces
        for word in words:
            # Emit each word
        storm.emit([word])
SplitSentenceBolt().run()
```

The "WordCount" class is responsible for counting the occurrences of each word. The code is as follows:

```java
public static class WordCount extends BaseBasicBolt
    {
    // Map to count occurrences of each word
     Map<String, Integer> counts = new HashMap<String,
        Integer>();

    @Override
        public void execute(Tuple tuple,
            BasicOutputCollector collector) {
        String word = tuple.getString(0); // Retrieve
            the first word from the Tuple
        Integer count = counts.get(word); // Get the
            current count for the word
        if (count == null) // If the word is not in the
            map, initialize the count to 0
        count = 0;
        count++; // Increment the count
        counts.put(word, count); // Update the count in
            the map
        collector.emit(new Values(word, count));      //
            Emit the word and its count
    }
    @Override
```

```
public void declareOutputFields(OutputFieldsDecla
  rer declarer) {
// Declare the fields of the stream for the next
  processing component
declarer.declare(new Fields("word", "count"));
  }
}
```

Exercises

(1) What are the three processes used in Storm's architecture?
(2) In Storm, what two types of topology components need to be constructed for each task implemented by the user?
(3) Briefly describe how to set up a Storm development environment.

Chapter 10

Spark: An In-Memory Big Data Computing Framework

This chapter covers Spark, an in-memory big data computing framework, including an overview of Spark, its execution mechanisms, execution modes, Spark RDDs, and the Spark ecosystem. Readers will gain an understanding of Spark's basic concepts, its operational principles, and task workflows. They will learn about Spark's three execution modes: Standalone mode, Spark YARN mode, and Spark Mesos mode. The chapter also explains the Resilient Distributed Dataset (RDD) programming model in Spark and the core components of the Spark ecosystem, including SparkSQL for structured data processing, Spark Streaming for real-time processing, GraphX for graph computations, and MLlib for machine learning.

10.1. Overview of Spark

Spark is a fast, general-purpose, and scalable big data computing framework developed by the AMP Lab at UC Berkeley in 2009. It is a general-purpose parallel framework similar to Hadoop MapReduce, with all the advantages of MapReduce. Unlike MapReduce, however, Spark stores intermediate results in memory, making it an in-memory big data computing framework. This design improves the real-time processing of data in big data environments while ensuring high fault tolerance and scalability. Spark allows users to deploy Spark clusters on inexpensive hardware. The Spark logo is shown in Figure 10.1.

Fig. 10.1.   Spark logo.

Spark was open-sourced in 2010, became an Apache incubator project in June 2013, and was promoted to a top-level Apache project in February 2014. The Spark ecosystem now includes several subprojects, such as SparkSQL, Spark Streaming, GraphX, and MLlib. It has gained popularity among numerous Internet companies, including IBM, Cloudera, Hortonworks, Baidu, Alibaba, Tencent, JD.com, and Youku. For example, Baidu uses Spark for big data search and direct navigation services; Alibaba utilizes GraphX to build large-scale graph computation and mining systems, implementing various recommendation algorithms; and Tencent's Spark cluster has reached 8,000 nodes, making it one of the largest Spark clusters in the world.

Spark is characterized by its speed, ease of use, versatility, and compatibility. Compared to Hadoop's MapReduce, Spark optimizes MapReduce performance significantly. Spark's in-memory computations can be more than 100 times faster than Hadoop's, and disk-based computations are also over 10 times faster than MapReduce. Spark implements an efficient Directed Acyclic Graph (DAG) execution engine, supporting high-performance data stream processing through in-memory computing. Programs can be written in Java, Scala, Python, or R, and Spark offers over 80 high-level operators to easily build parallel applications. Spark also supports interactive Python and Scala shells, allowing for convenient testing and problem-solving with Spark clusters. Additionally, Spark provides numerous libraries, such as MLlib for machine learning, GraphX for graph computation, and Spark Streaming for real-time stream processing, which can be seamlessly integrated within the same application. Spark can use Hadoop's YARN and Apache Mesos for resource management and scheduling, as well as its standalone cluster mode (Standalone mode) for built-in resource management and scheduling. It can access various data sources like HDFS, HBase, Cassandra, and S3.

Unlike Hadoop, which includes only MapReduce and HDFS, Spark's architecture comprises Spark Core and application frameworks built on

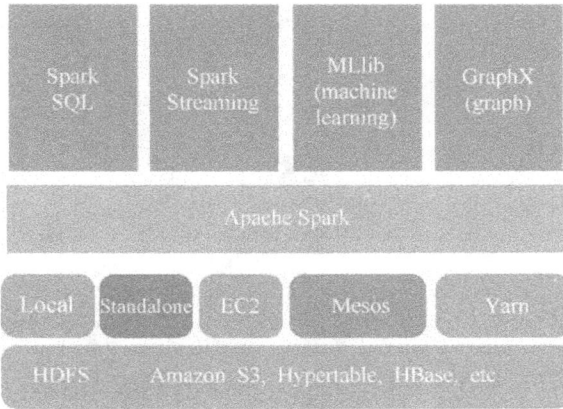

Fig. 10.2. Spark architecture.

Spark Core, including SparkSQL, Spark Streaming, MLlib, GraphX, Structured Streaming, and SparkR. Spark Core is the most critical component of Spark, analogous to MapReduce, handling offline data analysis. The Core library includes Spark's main entry points, such as the primary class used to write Spark programs, the application context (SparkContext), Resilient Distributed Datasets (RDDs), schedulers, shuffling and sorting of unstructured data, and serializers. SparkSQL provides an API for interacting with Spark via Hive Query Language (HiveQL), converting SparkSQL queries into Spark operations, with each database table treated as an RDD. Spark Streaming processes and controls real-time data streams, allowing real-time data to be handled similarly to regular RDDs. MLlib is Spark's library for machine learning algorithms, and GraphX offers algorithms and tools for graph control, parallel graph operations, and computations. The Spark architecture diagram is shown in Figure 10.2.

10.2. Spark's Execution Mechanism

(1) *Spark's Execution Architecture*: The execution architecture of Spark consists of four main components: the Cluster Manager, Worker Nodes that run job tasks, the Driver that controls the tasks of each application, and the Executors on each Worker Node responsible for executing specific tasks. The Spark execution architecture is illustrated in Figure 10.3.

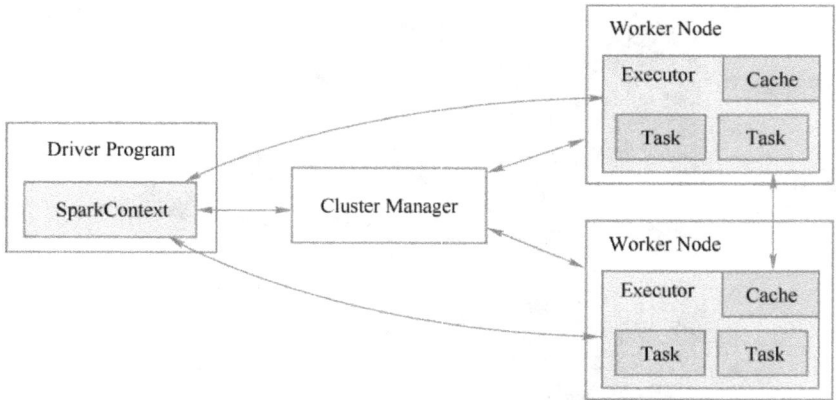

Fig. 10.3.   Spark's execution architecture.

The Driver is responsible for running the main function of the Spark application written by the user and for creating the SparkContext, which prepares the runtime environment for the application. The SparkContext communicates with the Cluster Manager to request resources, allocate and monitor tasks, start Executor processes, and send the application code and files to the Executors. Once the Executors complete their tasks, the Driver shuts down the SparkContext. The Cluster Manager can be Spark's built-in resource manager (Standalone mode), or it can be a resource manager like YARN or Mesos. A Worker Node refers to any node in the cluster capable of running application code. In Standalone mode, it is config-ured as a Worker node via the Slave file; in Spark YARN mode it corresponds to a NodeManager node; and in Spark Mesos mode, it corresponds to a Mesos Slave node. The Executor is a process running on a Worker Node, responsible for executing tasks (Tasks) and storing data either in memory or on disk.

Compared to the Hadoop MapReduce computing framework, Spark's use of the Executor process offers two advantages: First, the Executor utilizes multithreading to execute specific tasks (whereas Hadoop MapReduce uses separate processes for each task), thereby reducing the overhead of task initiation. Second, the Executor includes a BlockManager storage module that uses both memory and disk as storage devices. When multiple iterations of computation are needed, intermediate results can be stored in this storage module, allowing them to be accessed directly during subsequent iterations, without the need to read from or write to file systems like HDFS, effectively reducing I/O overhead. In interactive query scenarios, tables can be cached in this storage system in advance, improving read/write I/O performance.

From the above, it is clear that Spark's execution architecture has three main characteristics: Each application has its own dedicated Executor process, which persists for the duration of the application's runtime; Spark's execution process is independent of the resource manager, as long as it can obtain and maintain communication with the Executor process; tasks (Tasks) leverage optimizations such as data locality and speculative execution.

(2) *Basic Task Execution Flow in Spark*: The basic task execution flow in Spark is illustrated in Figure 10.4 and explained as follows:

- When a Spark application is submitted, a basic runtime environment needs to be set up for the application. This is done by the task control node (Driver) creating a SparkContext, which is responsible for communicating with the Cluster Manager,

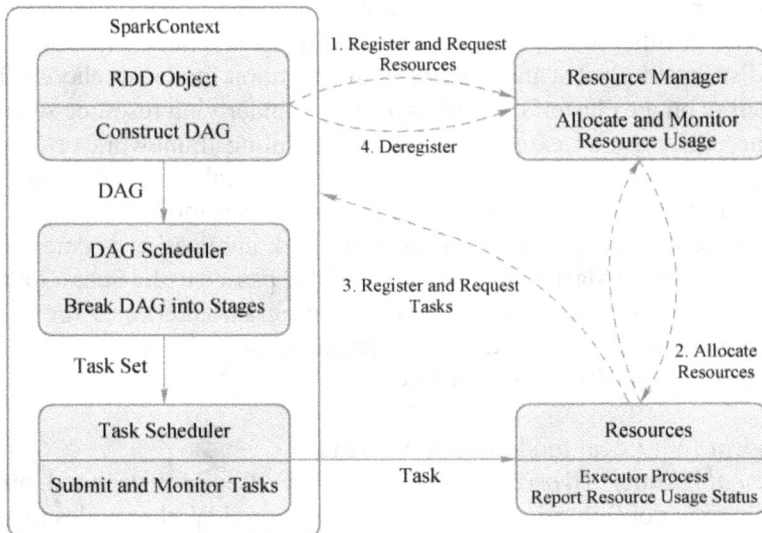

Fig. 10.4. Basic task execution flow in Spark.

requesting resources, allocating tasks, and monitoring execution. The SparkContext registers with the Cluster Manager and requests resources to run Executors.

- The Cluster Manager allocates resources for the Executors and starts the Executor processes, which report their status back to the Cluster Manager.
- The SparkContext constructs a DAG based on the RDD dependencies. The DAG is submitted to the DAG Scheduler for parsing, where it is broken down into multiple stages (each stage being a set of tasks), and the dependencies between stages are computed. These task sets are then handed over to the underlying TaskScheduler for execution. Executors request tasks from the SparkContext, which the TaskScheduler assigns to them, while the SparkContext distributes the application code to the Executors.
- The tasks are executed on the Executors, with the results being reported back to the TaskScheduler, which in turn reports to the DAG Scheduler. After the tasks are completed, the data are written, and all resources are released.

10.3. Spark's Execution Modes

Spark offers various execution modes. When deployed on a single machine, it can run in either local mode or pseudo-distributed mode. When deployed on a distributed cluster, there are several execution modes to choose from depending on the cluster's specific setup. The underlying resource scheduling can either use an external resource scheduling framework or Spark's built-in Standalone mode. Commonly used external resource scheduling frameworks include the YARN mode and the Mesos mode.

In practice, the execution mode of a Spark application is determined by the value of the Master environment variable passed to the SparkContext. Some modes may also require auxiliary program interfaces for proper operation. The supported Master environment variables are composed of the following specific strings or URLs:

(1) *Local [N]*: Local mode, using *N* threads.
(2) *Local Cluster [Worker, Core, Memory]*: Pseudo-distributed mode, allowing configuration of the number of virtual Worker nodes to start, as well as the number of CPUs and the amount of memory managed by each Worker node.

(3) *Spark://hostname:port*: Standalone mode, requiring Spark to be deployed on relevant nodes. The URL specifies the Spark Master host address and port.
(4) *Mesos://hostname:port*: Mesos mode, requiring both Spark and Mesos to be deployed on relevant nodes. The URL specifies the Mesos host address and port.
(5) *YARN Standalone/YARN Cluster*: One of the YARN modes, where both the main program logic and tasks run within the YARN cluster.
(6) *YARN Client*: Another YARN mode, where the main program logic runs locally, while specific tasks run within the YARN cluster.

10.3.1. *Standalone mode*

Standalone mode is Spark's built-in resource scheduling framework, consisting of three main node types: the Client node, the Master node, and the Worker nodes. The Driver can run either on the Master node or locally on the Client side. When a Spark job is submitted using the interactive "spark-shell" tool, the Driver runs on the Master node. However, when submitting a job using the "spark-submit" tool or running a Spark task on development platforms like Eclipse or IDEA using the "new SparkConf. setManager("spark://master:7077")" method, the Driver runs locally on the Client side. The execution process is illustrated in Figure 10.5.

(1) SparkContext connects to the Master, registers, and requests resources (CPU cores and memory).
(2) The Master determines which Worker to allocate resources based on the SparkContext's resource request and the information reported by Worker heartbeats.
(3) The Master identifies the Worker, allocates resources on that Worker, and starts the Standalone-ExecutorBackend (the daemon process for Executors in Standalone mode) on each Worker node.
(4) The StandaloneExecutorBackend registers with the SparkContext.
(5) On the Client node, SparkContext constructs a DAG based on the user program (completed within the RDD), decomposes the DAG into stages (TaskSets), and sends these stages to the TaskScheduler. The TaskScheduler dispatches the tasks to Executors on the allocated Workers for execution, specifically submitting them to the StandaloneExecutorBackend.

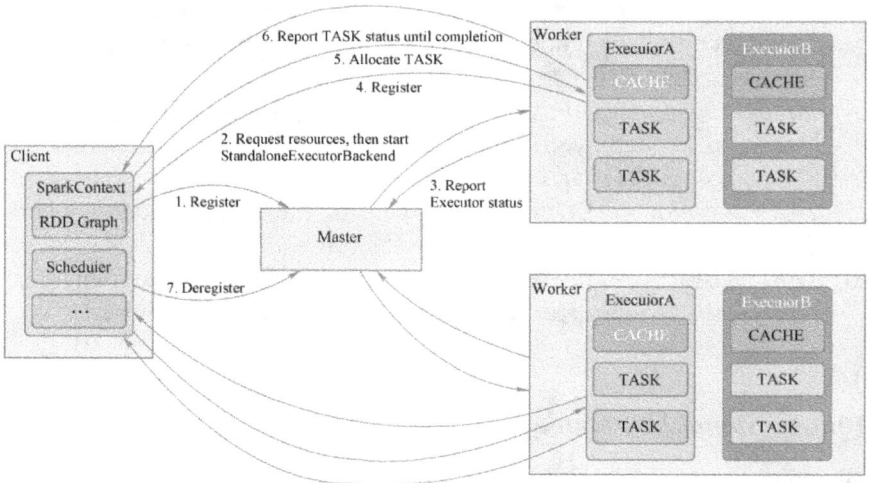

Fig. 10.5.　Standalone execution process.

(6)　The StandaloneExecutorBackend establishes an Executor thread pool, begins executing the tasks, and reports progress to the SparkContext until the tasks are completed.

(7)　Once all tasks are finished, SparkContext unregisters from the Master and releases the resources.

10.3.2.　*Spark on YARN mode*

Yet another Resource Negotiator (YARN) is a unified resource management mechanism that can run various computing frameworks. Many organizations use multiple frameworks besides Spark, such as MapReduce and Storm. To address this, Spark developed the Spark on YARN mode, leveraging YARN's robust resource management capabilities. This mode not only simplifies application deployment but also ensures complete isolation of resources between services and applications running on the YARN cluster. Additionally, YARN manages multiple services running simultaneously in the cluster through a queuing mechanism. Spark on YARN mode is classified into two types based on the Driver's location within the cluster: YARN-Client mode and YARN-Cluster mode (also referred to as YARN-Standalone mode). In production environments, the YARN-Cluster mode is generally used, while the YARN-Client mode is typically employed for interactive applications or debugging scenarios where immediate output is required.

(1) *YARN Framework Workflow*: Any framework that integrates with YARN must adhere to its development model. Before delving into the implementation details of Spark on YARN, it is necessary to understand some basic principles of the YARN framework. The fundamental workflow of the YARN framework is illustrated in Figure 10.6.

In the YARN framework, the ResourceManager is responsible for allocating cluster resources to various applications, with the fundamental unit of resource allocation and scheduling being the Container. Each Container encapsulates physical resources such as memory, CPU, disk, and network. Each task is allocated a Container and can only execute within that Container, utilizing the resources encapsulated by it. NodeManager, which is present on each computing node, is primarily responsible for launching Containers required by applications, monitoring resource usage (memory, CPU, disk, and network),

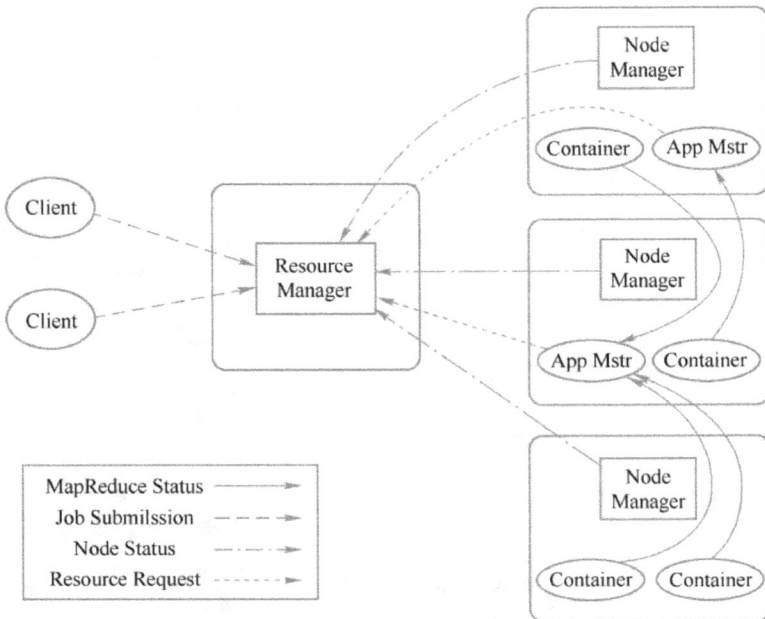

Fig. 10.6. Basic workflow of the YARN framework.

and reporting this information back to the ResourceManager. Together, ResourceManager and NodeManager form the core of the data computation framework. The ApplicationMaster (AppMstr) is associated with a specific application and is responsible for negotiating with the ResourceManager to acquire appropriate Containers, tracking their status, and monitoring their progress.

(2) *YARN-Cluster Mode*: In the YARN-Cluster mode, when a user submits an application to YARN, the application runs in two stages: The first stage involves launching the Spark Driver as an AppMstr in the YARN cluster; the second stage involves the AppMstr creating the application, requesting resources from the ResourceManager, and starting Executors to run Tasks, while monitoring the entire execution process until completion. The workflow of YARN-Cluster mode is detailed in Figure 10.7.

- *Submission*: Spark YARN-Client submits the application to the YARN cluster, including the AppMstr program, the command to launch the AppMstr, and the program to be run in the Executors.
- *Resource Allocation*: The ResourceManager receives the request and selects a NodeManager in the cluster to allocate the first

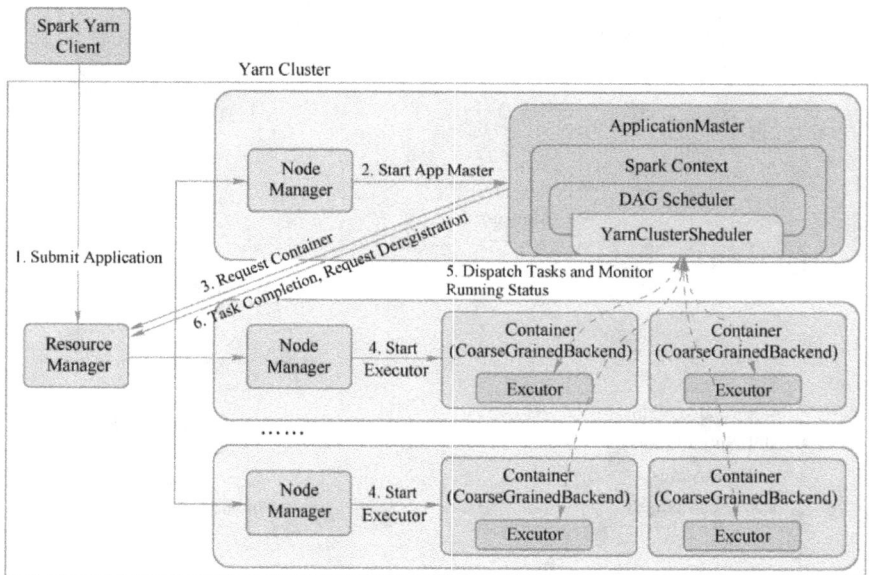

Fig. 10.7.   Workflow of YARN-Cluster mode.

Container for the application. The NodeManager is tasked with starting the AppMstr in this Container, where it initializes SparkContext and other components.

- *Registration and Resource Request*: The AppMstr registers with the ResourceManager, allowing users to view the application's status directly through the ResourceManager. The ResourceManager then uses a polling mechanism and RPC proto- col to request resources for various tasks and monitors their status until completion.

- *Executor Initialization*: Once the AppMstr acquires resources (i.e., Containers), it communicates with the corresponding NodeManager to start the CoarseGrainedBackend process (the Executor daemon) within the allocated Container. The CoarseGrainedBackend process then registers with the SparkContext in the AppMstr and requests Tasks. This process is similar to the Standalone mode, except that in Spark Application initialization, the CoarseGrainedBackend process works with the YARNClusterScheduler for task scheduling. The YARNClusterScheduler is a simple wrapper around TaskSchedulerImpl, adding functionalities like Executor waiting logic.

- *Task Execution*: The SparkContext within the AppMstr assigns Tasks to the CoarseGrainedBackend process. The CoarseGrainedBackend process executes the Tasks and reports the execution status and progress back to the AppMstr, allowing it to monitor the task status and restart tasks if they fail.

- *Completion*: After the application completes, the AppMstr requests to unregister and shut down from the ResourceManager.

(3) *YARN-Client Mode*: In the YARN-Client mode, the Driver runs locally on the client machine. This mode allows interaction between the Spark Application and the client, as the Driver is local and can be accessed via Web UI. By default, the Driver status can be accessed at "http://master_address:4040", while YARN can be accessed at "http://master_address:8088". The workflow of YARN-Client mode is out- lined in Figure 10.8.

(1) The Spark Yarn-Client requests the ResourceManager to initiate the AppMstr. During the SparkContext initialization, DAGScheduler and TASKScheduler are created. Since the Yarn-Client mode is chosen, the program selects

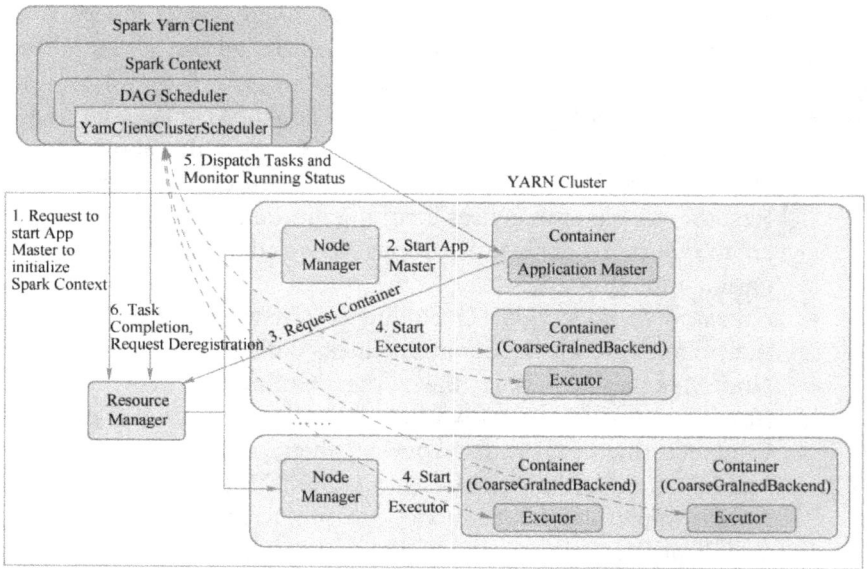

Fig. 10.8.   Yarn-Client mode workflow.

YarnClientClusterScheduler (running on the client side, whereas Yarn-Cluster runs within the resource manager) and YarnClientSchedulerBackend processes (also on the client side).

(2) Upon receiving the request, the ResourceManager selects a NodeManager within the cluster to allocate the first Container for the application, instructing it to start the AppMstr within this Container. Unlike Yarn-Cluster, the AppMstr does not run SparkContext but instead communicates with SparkContext to manage resource allocation.

(3) After the SparkContext in the Client is initialized, it establishes communication with the AppMstr and registers with the ResourceManager. Based on task information, it requests resources (Containers) from the ResourceManager.

(4) Once the AppMstr obtains the resources (Containers), it communicates with the respective NodeManager to start the CoarseGrainedBackend process within the allocated Containers. Upon starting, the CoarseGrainedBackend process registers with the SparkContext in the Client and requests tasks.

(5) The SparkContext in the Client assigns tasks to the CoarseGrainedExecutorBackend for execution. The CoarseGrainedExecutorBackend runs the tasks and reports their status and progress to the Driver, allowing the Client to monitor each task's state and restart any failed tasks if necessary.

(6) After the application completes, the SparkContext in the Client requests the ResourceManager to deregister and shuts down.

10.3.3. *Spark Mesos mode*

Spark can operate on a hardware cluster managed by Apache Mesos. The advantages of deploying Spark using Mesos include dynamic partitioning between Spark and other frameworks and the ability to scale partitions across multiple Spark instances. When a driver program creates a job and starts scheduling tasks, Mesos determines which machine handles each task. Multiple frameworks can coexist on the same cluster without relying on static resource allocation. If Mesos has only one master, the master's URL is "mesos://host:5050"; if Mesos has multiple masters (e.g., managed using Zookeeper), the master's URL is "mesos://zk://host1:2181,host2:2181,host3:2181/mesos".

Similar to Yarn, Mesos also offers two modes: Client and Cluster. In Client mode, the Spark Mesos framework is launched directly on the client machine and waits for the driver program's output. If the client shuts down, the driver stops running. In Mesos Cluster mode, the Driver program runs within the cluster, and shutting down the client does not affect the program's execution.

Spark can run in two modes on Mesos: coarse-grained mode and fine-grained mode. Coarse-grained mode is the default, while fine-grained mode has been deprecated since Spark 2.0. In coarse-grained mode, Mesos starts a long-running Spark task on each machine, and Spark tasks are dynamically scheduled as "mini-tasks" within it. This approach reduces startup latency but increases resource consumption, as Mesos must reserve resources for these long-running Spark tasks throughout their lifecycle. In fine-grained mode, each Spark task runs as an independent Mesos task, allowing multiple Spark instances (or other computing frameworks) to share machine resources more granularly. The resources available to each application increase or decrease as the application starts and stops, but each task's startup incurs a corresponding delay. This mode may not be suitable for low-latency scenarios such as interactive queries or web request responses.

10.4. Spark RDD

Spark provides a primary data abstraction known as RDDs, which is a concept of distributed memory. RDDs are collections distributed across nodes in a cluster and can be operated upon in parallel. RDDs feature automatic fault tolerance, location-aware scheduling, and scalability, allowing users to store data on disk and in memory while controlling data partitioning. Users can also request that Spark persist an RDD in memory for efficient reuse in parallel operations, and RDDs can automatically recover from node failures.

For developers, RDDs can be seen as Spark objects that operate in memory. For example, reading a file is an RDD, computing on the file is an RDD, and the result set is also an RDD. Different partitions, dependencies between data, and key-value type map data can all be considered RDDs.

10.4.1. *Characteristics of RDDs*

Data processing models typically fall into four categories: Iterative Algorithms, Relational Queries, MapReduce, and Stream Processing. Hadoop adopts the MapReduce model, while RDDs implement all four models, enabling Spark to be applied in various big data processing scenarios. Additionally, RDDs offer a rich set of operations for data computation. RDDs have the following five characteristics:

(1) *Partition*: The basic unit of data composition, RDDs provide a highly restricted shared memory model, where RDDs are essentially a collection of read-only partitioned records. An RDD consists of multiple partitions, and the size of the partitions determines the granularity of parallel computation. Each partition's computation is handled by an individual task. Users can specify the number of partitions when creating an RDD, with the default being the number of CPU cores allocated to the program.

(2) *Compute* (*Compute Function*): The compute function is the calculation function for each partition. In Spark, all computations are based on partitions, with each RDD achieving computation through the compute function.

(3) *Dependencies*: RDDs have dependencies on each other, which are primarily either narrow or wide. If each partition of the parent RDD can only be used by a single partition of the child RDD, this is called

a narrow dependency (e.g., map, filter, and union operations), as illustrated in Figure 10.9. If a partition of the parent RDD is used by multiple child RDD partitions, this is a wide dependency (e.g., groupByKey, reduceByKey, and sortByKey operations), as shown in Figure 10.10. When the number of partitions in the RDDs involved in a join operation is consistent, and the number of partitions in the resulting RDD is the same as that of the parent RDDs (joined with inputs co-partitioned), it is a narrow dependency. When each parent RDD partition corresponds to all child RDD partitions (joined with inputs not co-partitioned), it is a wide dependency.

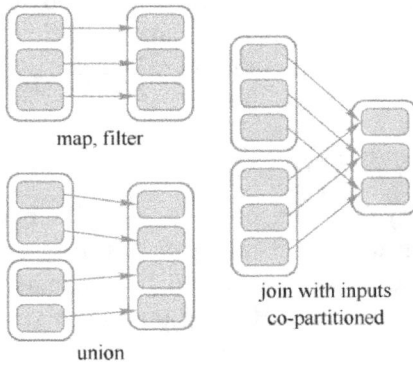

Fig. 10.9. Narrow dependencies.

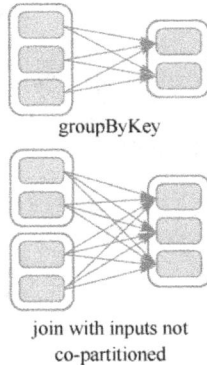

Fig. 10.10. Wide dependencies.

RDDs with narrow dependencies can be computed within the same stage, involving a shuffle process where all operations are executed together. Wide dependencies also involve a shuffle process, but the tasks of the subsequent RDD can only proceed after all tasks of the previous RDD have been completed.

(4) *Partitioner*: Spark currently implements two types of partitioner functions: HashPartitioner based on hashing and RangePartitioner based on ranges. The Partitioner is only applicable to key-value type RDDs, while non-key-value RDDs have a Partitioner value of none. The Partitioner function not only determines the number of partitions within the RDD itself but also dictates the number of partitions output after the parent RDD undergoes shuffling.

(5) *Preferred Locations*: Following the principle that "moving computation is better than moving data", Spark prioritizes task scheduling to the location where the data blocks are stored.

10.4.2. *Creation of RDDs*

There are two primary methods for creating RDDs: parallelizing an existing collection within the driver program, and referencing datasets from external storage systems such as shared file systems, HDFS, HBase, or any data source that provides a Hadoop InputFormat.

(1) *Parallelizing an Existing Collection within the Driver Program*: A parallel collection is created by invoking the "parallelize()" method of the "SparkContext" class (the main entry point for a Spark program) on an existing collection within the driver program (e.g., a Scala "Seq"). This process replicates the elements of the collection to form a distributed dataset that can be operated on in parallel. For example, you can create a collection with elements 1, 2, 3, 4, and 5 and perform parallel operations on it, as shown in the following:

```scala
scala> val data = Array(1,2,3,4,5) // Array containing
    elements 1, 2, 3, 4, 5
data: Array[Int] = Array(1, 2, 3, 4, 5) // The data
    type at this point is a collection (Array)
scala> val rdd1 = sc.parallelize(data) // Parallelizing
    the existing collection 'data'
```

```
rdd1: org.apache.spark.rdd.RDD[Int] =
ParallelCollectionRDD[2] at parallelize at
<console>:26 // The data type is now converted to
an RDD
```

One important parameter for parallel collections is the number of partitions into which the dataset is divided. Spark will run one task per partition in the cluster. Typically, each CPU in the cluster should handle 2–4 partitions. Spark attempts to automatically set the number of partitions based on the cluster configuration. However, you can also manually specify the number of partitions by passing it as the second parameter to the "parallelize" function. For example, "sc.parallelize(data, 10)" specifies that the dataset should be divided into 10 partitions.

(2) *Referencing Datasets from External Storage Systems*: Spark can create distributed datasets by reading from any storage source supported by Hadoop, including local file systems, HDFS, Cassandra, HBase, and Amazon S3. Additionally, Spark supports referencing text files, SequenceFiles (a file format used by Hadoop for storing binary key-value pairs), and any data source providing Hadoop InputFormat. Text files can be converted to RDDs by calling the "textFile()" method on "SparkContext". This method requires a file path (either a local path on the machine or paths like "hdfs://" or "s3n://") and reads the file as a collection of lines. For example, to create an RDD by reading the "movies.dat" file located at "/home/student/data", you would use the following command:

```
scala> val rdd2 = sc.textFile("/home/student/data/
movies.dat") // Read file
rdd2:org.apache.spark.rdd.RDD[String]=/home/student/
data/movies.dat MapPartitionsRDD[4] at textFile at
<console>:24 // Data type converted to RDD
```

The process of reading files with Spark is as follows:

(1) For local file systems, the file to be read must be accessible via the same path on all Worker nodes. Therefore, either copy the file to the same path on all Worker nodes or mount a shared file system.
(2) All Spark file-based input methods (including "textFile") support input parameters such as directories, compressed files, and wildcards,

for example, "textFile("/my/directory")", "textFile("/my/directory/*.txt")" and "textFile("/my/directory/*.gz")".

(3) The "textFile()" method also accepts an optional second parameter to control the number of file partitions. By default, Spark creates one partition for each block of the file (with the default HDFS block size of 128 MB). This parameter can be adjusted to control the number of data partitions. Note that the number of partitions cannot be less than the number of blocks. Besides text files, Spark's Scala API supports other data formats as described in the following:

- "SparkContext.wholeTextFiles" can read directories containing multiple small text files and returns the result as (filename, content) pairs. Unlike "textFile", which returns only the content of the files with each line as a record, "wholeTextFiles" returns the content as (filename, content) pairs. The partitioning is determined by the data location, which may sometimes lead to too few partitions. In such cases, "wholeTextFiles" offers an optional second parameter to control the minimum number of partitions.

- For "SequenceFiles", use "SparkContext.sequenceFile[K, V]", where K and V are the types of keys and values in the file, respectively. These types should be subclasses of the "Writable" interface (which is a simple and efficient serialization object based on "DataInput" and "DataOutput"), such as "IntWritable" and "Text". Additionally, Spark allows specifying native types for some common "Writable" types, such as "sequenceFile[Int, String]" which automatically reads "IntWritable" and "Text".

- For other Hadoop "InputFormat" types, use "SparkContext.hadoopRDD()" with any "JobConf" object, "InputFormat", key class, and value class. This is similar to setting the input source for a Hadoop Job. Alternatively, "SparkContext.newAPIHadoopRDD" can be used, which accepts an "InputFormat" based on the newer Hadoop MapReduce API (org.apache.hadoop.mapreduce) as a parameter.

(4) "RDD.saveAsObjectFile" and "SparkContext.objectFile" support saving RDD elements in a serialized Java object format. Although this serialization method is less efficient than Avro, it provides a convenient way to save RDDs.

10.4.3. *Basic RDD operations*

RDDs support two types of operations: transformations and actions. Transformation operations create a new dataset from an existing one, while action operations perform computations on the dataset and return a value to the driver or write the results to an external system, thereby triggering actual computation. For example, "Map" is a transformation operation that applies a function to each element of the dataset and returns a new RDD representing the result. "Reduce" is an action operation that aggregates all elements of an RDD using a given function and returns the final result to the driver. The type of the return value can be used to determine whether a function is a transformation or an action: Transformations return an RDD, while actions return a specific data type.

All transformations in Spark are lazy; they do not compute their results immediately. Transformations are only computed when an action requires returning a result to the driver. This design enables Spark to run more efficiently. For instance, if a dataset created by "Map" is used in "Reduce", the transformation will only return the reduced result to the driver, rather than the larger mapped dataset. The type of the return value can be used to distinguish between transformations and actions, with transformations returning an RDD and actions returning a specific data type.

By default, each time an operation is performed on an RDD, each transformed RDD can be recomputed. However, you can use the "persist" (or "cache") method to persist an RDD in memory, allowing Spark to access the elements faster in subsequent queries. Transformations also support persisting RDDs on disk or replicating them across multiple nodes. Commonly used transformation operations in Spark are shown in Table 10.1, and commonly used action operations are shown in Table 10.2.

10.4.4. *RDD persistence (cache)*

One of the most important features of Spark is the ability to persist (or cache) datasets in memory. After an RDD is persisted, each node will store the computed partition results in memory and reuse them in subsequent operations on that dataset or its derivatives. Persistence speeds up the execution of subsequent operations (often by more than tenfold) and is a crucial tool for iterative algorithms and fast interactive queries. You can persist an RDD using either the "persist()" method or the "cache()"

Table 10.1.   Common RDD transformation operations.

Transformation Operation	Meaning
map(*func*)	Returns a new RDD where each element from the source dataset is transformed by the "func" function.
filter(*func*)	Returns a new RDD containing only the elements for which the "func" function returns "true".
flatMap(*func*)	Similar to "map", but each input item can be mapped to zero or more output items, so the "func" function should return a sequence (Seq) instead of a single item.
mapPartitions(*func*)	Similar to "map" but runs independently on each partition (block) of the RDD. Therefore, when running on an RDD of type T, the "func" function must be of type "Iterator<T> => Iterator<U>".
mapPartitionsWithIndex(*func*)	Similar to "mapPartitions", but the "func" function takes an additional integer parameter representing the partition index. Therefore, when running on an RDD of type T, the "func" function must be of type "(Int, Iterator<T>) => Iterator<U>".
sample(*withReplacement, fraction, seed*)	Samples the data according to the fraction specified, with the option to replace using random numbers. The "seed" parameter specifies the seed for the random number generator.
union(*otherDataset*)	Returns a new RDD that is the union of the source dataset and the elements in the parameter.
intersection(*otherDataset*)	Returns a new RDD that is the intersection of the source dataset and the elements in the parameter.
distinct([*numTasks*])	Returns a new RDD with duplicates removed from the source dataset.

Table 10.1. (*Continued*)

Transformation Operation	Meaning
groupByKey([*numTasks*])	When called on an RDD of type (K,V), returns an RDD of type (K, Iterable<V>). For aggregation operations like sum or average, it is better to use "reduceByKey" or "aggregateByKey". By default, the output's parallelism depends on the number of partitions in the parent RDD, but it can be set to a different number of tasks using an optional "numTasks" parameter.
reduceByKey(*func*, [*numTasks*])	When called on an RDD of type (K,V), returns an RDD of type (K,V) by aggregating the values with the same key using the specified "reduce" function. Similar to "groupByKey", the number of reduce tasks can be set using an optional "numTasks" parameter.
aggregateByKey(*zeroValue*) (*seqOp, combOp,* [*numTasks*])	When called on an RDD of type (K,V), returns an RDD of type (K,U). Aggregates values with the same key using the given combine function and a neutral initial value ("zeroValue"). The return type allows for a different type than the input value.
sortByKey([*ascending*], [*numTasks*])	When called on an RDD of type (K,V), where K must implement the Ordered interface, returns an RDD of type (K,V) sorted by key.
join(*otherDataset*, [*numTasks*])	When called on RDDs of type (K,V) and (K,W), returns an RDD of type (K, (V, W)) where all elements with the same key are paired together. Outer joins are also supported with "leftOuterJoin", "rightOuterJoin", and "fullOuterJoin".
cogroup(*otherDataset*, [*numTasks*])	When called on RDDs of type (K,V) and (K,W), returns an RDD of type (K, (Iterable<V>, Iterable<W>)). This operation is also known as "groupWith".
cartesian(*otherDataset*)	When called on RDDs of type T and U, returns an RDD of type (T,U), i.e., the Cartesian product.

(*Continued*)

Table 10.1.   (*Continued*)

Transformation Operation	Meaning
pipe(*command, [envVars]*)	Connects each partition of the RDD through a shell command. RDD elements are written to the process's stdin, and the lines of output from stdout are returned as a string RDD.
coalesce(*numPartitions*)	Reduces the number of partitions in the RDD to the specified "numPartitions", effectively filtering large datasets.
Repartition(*numPartitions*)	Reshuffles the RDD data and redistributes it randomly across new partitions, leading to a more balanced data distribution. The number of new partitions depends on "numPartitions". This method requires shuffling all data across the network.
repartitionAndSortWithinPartitions (*partitioner*)	Repartitions the RDD according to the specified "partitioner" and sorts the data by key within each resulting partition. This is a composite operation that is functionally equivalent to first repartitioning and then sorting within each partition, but it is optimized internally (pushing the sorting process into the shuffle), resulting in better performance.

method. However, these methods do not immediately cache the RDD; it is only stored in memory on the node during the first computation in an operation, making it available for reuse in subsequent operations. Spark's caching mechanism is fault-tolerant—if any partitions of the RDD are lost, they will be automatically recomputed using the original transformations that created them.

Additionally, each persisted RDD can be stored with different storage levels, such as saving datasets on disk, storing them in memory as serialized Java objects, or replicating them across nodes. These storage levels are set by passing a "StorageLevel" object to the "persist()" method. The "cache()" method uses the default storage level: "StorageLevel. MEMORY_ONLY" (stores deserialized objects in memory). The storage levels are detailed in Table 10.3.

Table 10.2. Common RDD action operations.

Action Operation	Meaning
reduce(*func*)	Aggregates all elements in the RDD using the "func" function, which must be both commutative and associative.
collect()	Returns all elements of the RDD as an array to the driver program.
count()	Returns the number of elements in the RDD.
first()	Returns the first element of the RDD (similar to "take(1)").
take(*n*)	Returns an array containing the first "n" elements of the RDD.
takeSample(*withReplacement*, *num*, [*seed*])	Returns an array of "num" elements sampled randomly from the RDD, with the option to replace missing parts using random numbers. The "seed" parameter specifies the seed for the random number generator.
takeOrdered(*n*, [*ordering*])	Returns the first "n" elements of the RDD using natural ordering or a custom comparator.
saveAsTextFile(*path*)	Writes the elements of the dataset as a text file (or a set of text files) to the local file system, HDFS, or any other file system supported by Hadoop. Spark calls the "toString" method on each element to convert it into a line of text in the file.
saveAsSequenceFile(*path*) (Java and Scala)	Writes the elements of the dataset to a Hadoop SequenceFile at the given path in the local file system, HDFS, or any other file system supported by Hadoop.
saveAsObjectFile(*path*) (Java and Scala)	Uses Java serialization to write the elements of the dataset in a simple format, which can then be loaded using the "SparkContext.objectFile()" method.
countByKey()	For an RDD of type (K,V), returns a hashmap of type (K, Int), representing the number of elements corresponding to each key.
foreach(*func*)	Runs the "func" function on each element of the dataset to perform updates.

Table 10.3. Storage levels.

Storage Level	Description
MEMORY_ONLY	Default level. Stores the RDD as deserialized Java objects in the JVM. If the RDD does not fit in memory, some partitions will not be cached and will be recomputed when needed.
MEMORY_AND_DISK	Stores the RDD as deserialized Java objects in the JVM. If the RDD does not fit in memory, partitions are stored on disk and read from there when needed.
MEMORY_ONLY_SER (Java and Scala)	Stores the RDD as serialized Java objects (one byte array per partition). This typically saves space compared to deserialized storage, especially with a fast serializer, but intense read operations will consume more CPU resources.
MEMORY_AND_DISK_SER (Java and Scala)	Similar to "MEMORY_ONLY_SER", but partitions that do not fit in memory are stored on disk instead of being recomputed each time they are needed.
DISK_ONLY	Stores RDD partitions only on disk.
MEMORY_ONLY_2, MEMORY_AND_DISK_2, etc	Same as the levels above, but each partition is replicated on two cluster nodes.
OFF_HEAP (experimental)	Similar to "MEMORY_ONLY_SER" but stores data in off-heap memory. This requires off-heap memory to be enabled.

The storage levels in Spark allow developers to balance memory usage and CPU efficiency. The following guidelines are recommended for selecting an appropriate storage level:

(1) If the RDD fits within the default storage level ("MEMORY_ONLY"), choose the default level. This is the most CPU-efficient option and ensures that operations on the RDD run as quickly as possible.
(2) If the RDD does not fit the default level, try using "MEMORY_ONLY_SER" and select a fast serialization library to improve space efficiency while still allowing quick access.
(3) Avoid storing the RDD on disk unless the time spent computing the RDD is significant or a large amount of data needs to be filtered.

Otherwise, recomputing a partition can be as slow as reading data from disk.

(4) If rapid fault recovery is required, such as when using Spark to respond to requests from a web application, use a replication storage level. All storage levels support full fault tolerance by recomputing lost data, but replicated data allow tasks to continue operating on the RDD without needing to recompute lost partitions.

Spark automatically monitors cache usage on each node. If the data to be cached exceed available memory, older data partitions are evicted using a Least Recently Used (LRU) caching strategy. Additionally, you can manually remove persisted RDDs from the cache using the "RDD. unpersist()" method.

10.4.5. *Spark shared variables*

Another important abstraction in Spark is the shared variable, which can be used in parallel operations. By default, when Spark executes a set of tasks in parallel across different nodes, it sends a copy of each variable to every task. However, there are cases where variables need to be shared between tasks or between tasks and the driver. Typically, when you pass a function to a Spark operation (such as "map" or "reduce") that is executed on remote cluster nodes, a copy of all variables within the function is used. These variables are copied to each machine, and any updates to these variables on remote machines are not sent back to the driver. Using general read-write shared variables between tasks is inefficient. However, Spark provides two limited types of shared variables for common use cases: broadcast variables and accumulators.

(1) *Broadcast Variables*: Broadcast variables are used to cache values in memory across all nodes, allowing programmers to keep a read-only variable on each machine instead of copying it to each task. Broadcast variables can be used to efficiently provide each node with a copy of a large input dataset. Spark also attempts to use an efficient broadcast algorithm to distribute broadcast variables, reducing communication costs.

Broadcast variables are created by calling the "SparkContext. broadcast(v)" method on a variable "v". A broadcast variable is a wrapper around the variable "v" and can be accessed by calling the "value" method, as shown in the following:

```scala
scala> val broadcastVar = sc.broadcast(Array(1, 2,
  3))
broadcastVar:org.apache.spark.broadcast.
  Broadcast[Array[Int]] = Broadcast(0)
scala> broadcastVar.value
res0: Array[Int] = Array(1, 2, 3)
```

After creating a broadcast variable, it should be used in all functions across the cluster in place of the variable "v", ensuring that "v" is not transmitted between nodes more than once. Additionally, to guarantee that all nodes receive the same value of the broadcast variable, the object "v" should not be modified after it has been broadcast (especially if the variable might later be sent to a new node).

(2) *Accumulators*: Accumulators are another type of variable that is broadcast to worker nodes. Unlike broadcast variables, which are read-only, accumulators allow for accumulation. Accumulators provide a simple syntax for aggregating values from worker nodes back to the driver program. They are efficiently used in parallel operations for tasks such as counting or summing. Global accumulators are accessible only by the driver program, and each worker node can only access and operate on its local accumulator. Similarly, the value of an accumulator is accessed via the value method. Native Spark supports accumulators for numeric types only, although developers can extend support to new types.

As a user, you can create named or unnamed accumulators. For instance, as shown in Figure 10.11, a named accumulator, such as counter, can be displayed on Spark's web interface during the stage that modifies the accumulator. Additionally, the Spark task UI will show the values of each accumulator modified by the tasks.

Accumulators are only updated within action operations, and Spark ensures that each task's updates to an accumulator are applied only once, meaning that the value will not be updated even if the task is restarted. During transformation operations, when a task or stage is re-executed, each task may update the accumulator more than once.

Accumulators do not alter Spark's lazy evaluation model. If they are updated during an RDD operation, their value is only updated when an

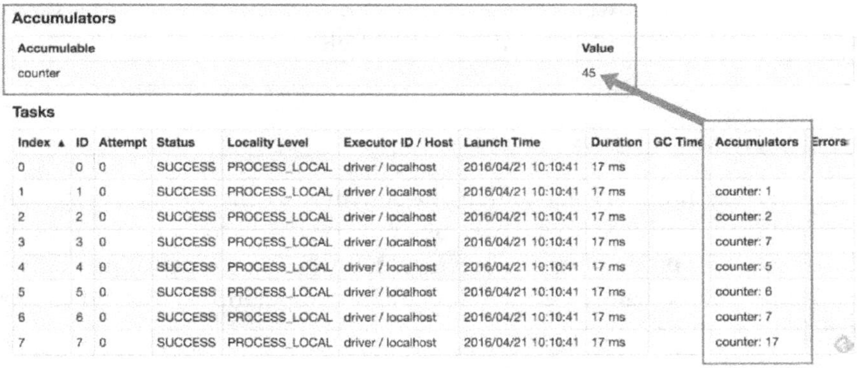

Accumulators

Accumulable	Value
counter	45

Tasks

Index ▲	ID	Attempt	Status	Locality Level	Executor ID / Host	Launch Time	Duration	GC Time	Accumulators	Errors
0	0	0	SUCCESS	PROCESS_LOCAL	driver / localhost	2016/04/21 10:10:41	17 ms			
1	1	0	SUCCESS	PROCESS_LOCAL	driver / localhost	2016/04/21 10:10:41	17 ms		counter: 1	
2	2	0	SUCCESS	PROCESS_LOCAL	driver / localhost	2016/04/21 10:10:41	17 ms		counter: 2	
3	3	0	SUCCESS	PROCESS_LOCAL	driver / localhost	2016/04/21 10:10:41	17 ms		counter: 7	
4	4	0	SUCCESS	PROCESS_LOCAL	driver / localhost	2016/04/21 10:10:41	17 ms		counter: 5	
5	5	0	SUCCESS	PROCESS_LOCAL	driver / localhost	2016/04/21 10:10:41	17 ms		counter: 6	
6	6	0	SUCCESS	PROCESS_LOCAL	driver / localhost	2016/04/21 10:10:41	17 ms		counter: 7	
7	7	0	SUCCESS	PROCESS_LOCAL	driver / localhost	2016/04/21 10:10:41	17 ms		counter: 17	

Fig. 10.11.   Display of accumulators in Spark's web interface.

action operation is performed on the RDD. Therefore, updates to accumulators in lazy transformations such as "map()" are not guaranteed to be actually executed. The following code snippet demonstrates this property:

```
val accum = sc.longAccumulator
data.map { x => accum.add(x); x }
// The value of accum remains 0 because no action
    operation has triggered the actual computation of
    the map
```

In summary, shared variables in Spark can be used to perform global operations, such as updating the total count of records or broadcasting some large configuration variables. For example, broadcast variables can be used to monitor changes in a database, enabling the periodic rebroadcasting of new table configurations.

```
myVector.add(v)
{

}
// Create an accumulator of the following type
val myVectorAcc = new VectorAccumulatorV2
// Register it with the Spark context
sc.register(myVectorAcc, "MyVectorAcc1")
```

Please note that when developers define their own accumulator types, the result type may differ from the type of elements being added.

10.5. Spark Ecosystem

This section provides a detailed introduction to the Spark ecosystem, focusing primarily on four core sub-frameworks based on Spark Core: Spark SQL for processing structured data, Spark Streaming for processing real-time data streams, GraphX for graph computations, and MLlib for machine learning algorithms.

10.5.1. *Spark SQL*

The first query engine used by Spark was Hive, provided by Hadoop. Since Hive's underlying engine is based on MapReduce, which is disk-based, its performance was extremely low. Later, Spark introduced Shark, which was still closely associated with Hive, relying heavily on Hive for many underlying functions. However, Shark modified the memory management, physical planning, and execution modules, using Spark's in-memory computation model at its core, thereby significantly improving performance over Hive. However, since Shark's underlying architecture depended on Hive's syntax parser, query optimizer, and other components, there were certain limitations on its performance improvement. Consequently, the Spark team decided to completely abandon Shark and introduced Spark SQL starting from Spark 1.0.

Spark SQL is a module in Spark used for processing structured data. Unlike the basic API of Spark RDD, the Spark SQL interface provides more information about the data structure and the computation being performed. Internally, Spark SQL uses this information to optimize performance. Spark SQL can be interacted with via SQL or the Dataset API. When computations are performed by the same engine, different APIs and languages are available for use. This unification allows developers to switch seamlessly between the most familiar APIs to complete the same computational tasks. Spark SQL has the following features:

(1) *Easy Integration*: Seamlessly integrates SQL queries with Spark programs. Spark SQL allows querying structured data in a Spark program

using SQL or the familiar DataFrame API. It supports Java, Scala, Python, and R languages.

(2) *Unified Data Access*: Connects to any data source in the same way. DataFrame and SQL provide common methods for accessing various data sources, including Hive, Avro, Parquet, ORC, JSON, and JDBC.

(3) *Hive Compatibility*: Can run SQL or HiveQL queries on existing Hive warehouses. Spark SQL supports HiveQL syntax as well as Hive SerDes and UDFs, allowing access to existing Hive warehouses.

(4) *Standard Data Connectivity*: Connects via JDBC or ODBC. Supports external tools such as business intelligence software to connect to Spark SQL for querying through standard database connectors (JDBC/ODBC).

From the above, it is evident that Spark's structured data processing mainly includes Spark SQL, DataFrame, Dataset, and related components.

A Dataset is a distributed collection of data and was introduced as a new interface in Spark 1.6. It combines the benefits of RDDs with the optimization advantages of Spark SQL's execution engine. A Dataset can be constructed through the Java Virtual Machine (JVM) and then manipulated using transformation operations such as map, flatMap, and filter. The Dataset API is available in both Java and Scala.

A DataFrame is a Dataset organized into named columns. Conceptually, it is equivalent to a table in a relational database or a data frame in R/ Python but with more optimizations at the underlying level. A DataFrame can be constructed from a wide range of data sources, such as structured data files, Hive tables, various databases, or existing RDDs. The DataFrame API is supported in Java, Python, Scala, and R. In Scala and Java, a DataFrame is represented by a Dataset of Rows; in the Scala API, a DataFrame can be simply considered as a type alias for Dataset[Row]; in the Java API, users need to use Dataset<Row> to represent a DataFrame.

10.5.2. *Spark streaming*

Spark Streaming is designed to address stream processing challenges and is an extension of the core Spark API. It supports scalable, high-throughput, and fault-tolerant processing of real-time data streams and can be seamlessly integrated with other Spark modules. Spark Streaming supports data acquisition from various sources such as Kafka, Flume, Kinesis, and HDFS. Once the data are acquired, they can be processed using advanced functions like map, reduce, join, and window. Finally, the processed results can be pushed to file systems, databases, and other destinations, as illustrated in Figure 10.12. Additionally, Spark Streaming can seamlessly integrate with MLlib (machine learning) and GraphX.

Spark Streaming is a coarse-grained framework, meaning it can only apply specified processing methods to a batch of data. Its core data processing is based on a micro-batch architecture. The internal workflow of Spark Streaming is illustrated in Figure 10.13. After Spark Streaming is started, data continuously flow in through the input data stream and are divided into different jobs (i.e., batches of input data) based on time. Spark Streaming receives real-time data streams and breaks the data into batches for processing by the Spark Engine, which then produces the final output stream in batches.

Spark Streaming provides a high-level abstraction called DStream, which represents a continuous stream of data as well as the resulting data stream after applying various Spark primitives (Spark primitives refer to Transformation and Action operations). Internally, in Spark Streaming, a DStream is represented by a series of continuous RDDs, where each RDD contains data from a specific time interval, as shown in Figure 10.14. A DStream can be created from an external input source, or new DStreams can be generated by performing transformations on existing DStreams.

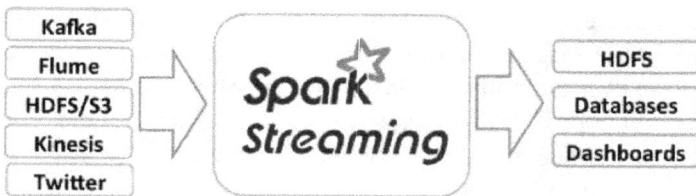

Fig. 10.12.   Spark Streaming architecture.

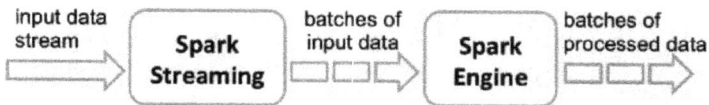

Fig. 10.13.   Spark Streaming internal workflow.

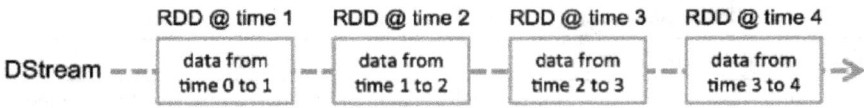

Fig. 10.14. DStream workflow.

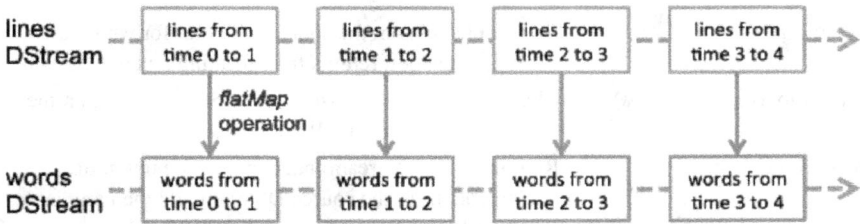

Fig. 10.15. DStream processing workflow.

Data operations are also performed on a per-RDD basis. After the stream data are divided into batches, a first-in, first-out queue is created, from which the Spark Engine sequentially retrieves each batch of data, encapsulates each batch as an RDD, and processes it, as illustrated in Figure 10.15. Suppose "lines" is a DStream divided into four batches by time intervals, with data stored in four consecutive RDDs (e.g., "lines from time 0 to 1", "lines from time 1 to 2"). Performing a "flatMap" operation on the "lines" DStream effectively applies the "flatMap" operation to each RDD within the DStream, transforming the "lines" DStream into a "words" DStream. The resulting data are stored in the RDDs corresponding to the "words" DStream, which are composed of specific time intervals (e.g., "words from time 0 to 1", "words from time 1 to 2").

Similar to RDDs, DStreams include two types of operations: transformation operations and output operations. The standard transformation operations and window transformation operations discussed as follows fall under transformation operations:

(1) *Standard Transformation Operations*: The standard transformation operations for DStreams are shown in Table 10.4.
(2) *Window Transformation Operations*: Spark Streaming also provides window calculations, allowing data transformations through sliding windows, as shown in Figure 10.16. Each time the window slides over

Table 10.4.   Standard transformation operations for DStream.

Transformation Operation	Meaning
map(*func*)	Passes each element of the source DStream through the function "func", returning a new DStream.
flatMap(*func*)	Similar to "map", but each input item can be mapped to 0 or more output items.
filter(*func*)	Selects elements in the source DStream for which the "func" function returns true, forming a new DStream.
repartition(*numPartitions*)	Changes the partition size of the DStream based on the input parameter "numPartitions".
union(*otherStream*)	Returns a new DStream that contains the union of elements from the source DStream and the parameter "otherStream".
count()	Counts the number of elements in each RDD of the source DStream, returning a DStream with a single RDD containing one element.
reduce(*func*)	Aggregates the elements of each RDD in the source DStream using the function "func", returning a DStream with a single RDD containing one element. The "func" function must be commutative and associative.
countByValue()	Counts the frequency of each element in every RDD of the DStream and returns a new DStream of type "[(K, Long)]", where "K" is the type of element in the RDD and "Long" is the frequency of occurrence.
reduceByKey(*func*, [*numTasks*])	When called on a DStream of type "(K, V)", returns a new DStream of type "(K, V)", where the value "V" for each key is aggregated using "func". Note: By default, tasks are submitted with Spark's default parallelism (two in local mode, and determined by the "spark.default.parallelism" configuration property in cluster mode). The number of parallel tasks can be set with "numTasks".
join(*otherStream*, [*numTasks*])	When called on DStreams of type "(K, V)" and "(K, W)", returns a new DStream of type "(K, (V, W))".
cogroup(*otherStream*, [*numTasks*])	When called on DStreams of type "(K, V)" and "(K, W)", returns a new DStream of type "(K, Seq[V], Seq[W])".

Table 10.4. (*Continued*)

Transformation Operation	Meaning
transform(*func*)	Returns a new DStream by applying an RDD-to-RDD function to each RDD in the source DStream, allowing arbitrary RDD operations on the DStream.
updateStateByKey(*func*)	Returns a new state DStream where each key's state is updated based on the previous state and the new value for the key using the given function "func". This can be used to maintain arbitrary state data for each key.

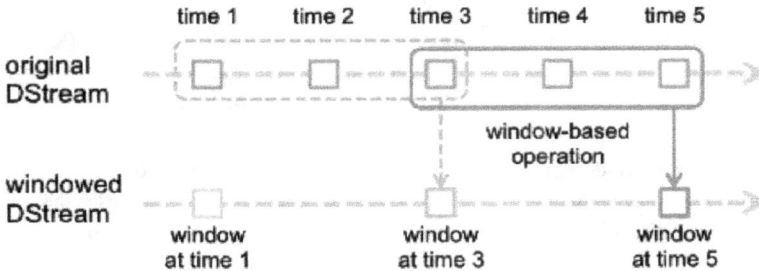

Fig. 10.16. Window transformation workflow.

the source DStream, the source RDDs within the window are combined to produce a DStream for the window. This operation is applied to the last 3 time units of data, with a sliding interval of 2 time units. This indicates that any window operation requires specifying two parameters: window length (the duration of the window, 3 in the figure) and sliding interval (the interval at which the window operation is performed, 2 in the figure). These two parameters must be multiples of the source DStream's batch interval (1 in the figure). Common window operations are listed in Table 10.5.

(3) *Output Operations*: Spark Streaming allows data from a DStream to be output to external systems, such as databases or file systems. Output operations essentially make the data resulting from transformation operations available for use via external systems while also triggering the actual execution of all DStream transformation operations (similar to RDD operations). The main output operations for DStream are shown in Table 10.6.

Table 10.5. Common window transformation operations.

Window Transformation Operation	Meaning
window(*windowLength, slideInterval*)	Returns a new DStream based on window batch computations from the source DStream.
countByWindow(*windowLength, slideInterval*)	Returns the count of elements in the DStream based on the sliding window.
reduceByWindow(*func, windowLength, slideInterval*)	Applies the function "func" to aggregate elements from the source DStream within the sliding window, returning a DStream with a single RDD containing one element. The "func" function must be commutative and associative.
reduceByKeyAndWindow(*func, windowLength, slideInterval, [numTasks]*)	Aggregates values in a DStream of type (K, V) by key K using the function "func" based on the sliding window, returning a new DStream of type (K, V).
reduceByKeyAndWindow(*func, invFunc, windowLength, slideInterval, [numTasks]*)	A more efficient implementation of "reduceByKeyAndWindow" that uses the previous window's reduced values to incrementally compute the reduce values for each window. This is achieved by reducing new data entering the sliding window and "reverse reducing" old data from the expired window.
countByValueAndWindow(*windowLength, slideInterval, [numTasks]*)	When called on a DStream of type (K, V), calculates the frequency of each element within each RDD of the source DStream based on the sliding window, returning a DStream of type (K, Long), where K represents the type of element in the RDD and Long represents the frequency of occurrences. The optional parameter "numTasks" can configure the number of reduce tasks.

(4) *Comparison between Spark Streaming and Storm*:

- *Processing Model and Latency*: Both Spark Streaming and Storm frameworks offer scalability and fault tolerance, but their processing models are fundamentally different. Spark Streaming processes multiple events (batches) within a short time window, while Storm achieves sub-second latency by processing one event

Table 10.6. DStream output operations.

Output Operation	Meaning
print()	Prints the first 10 elements of the DStream data running on the application driver node. This is useful for development and debugging.
saveAsTextFiles(*prefix*, [*suffix*])	Saves the content of the DStream as text files, with files generated during each batch interval named in the format "prefix-TIME_IN_MS[.suffix]".
saveAsObjectFiles(*prefix*, [*suffix*])	Saves the content of the DStream serialized as objects in SequenceFile format, with files generated during each batch interval named in the format "prefix-TIME_IN_MS[.suffix]".
saveAsHadoopFiles(*prefix*, [*suffix*])	Saves the content of the DStream as Hadoop files, with files generated during each batch interval named in the format "prefix-TIME_IN_MS[.suffix]".
foreachRDD(*func*)	The most general output operation, which applies the function "func" to the RDDs in the DStream. This operation outputs data to external systems, such as saving RDDs to files or network databases. Note that the "func" function is executed in the Driver process running the streaming application.

at a time. Therefore, Storm can achieve sub-second latency, while Spark Streaming has some inherent latency.

- *Fault Tolerance and Data Guarantees*: Both Spark Streaming and Storm provide data guarantees during fault tolerance, but Spark Streaming offers better support for stateful computations. In Storm, every record must be tracked and marked throughout its movement in the system, so Storm can only guarantee that each record is processed at least once, but it allows for multiple processing in the case of recovery from errors. This means mutable states might be updated multiple times, potentially leading to incorrect results. On the other hand, Spark Streaming only needs to track records at the batch processing level, ensuring that each batch record is processed exactly once.
- *Implementation and Programming API*: Spark Streaming is implemented in Scala, while Storm is primarily implemented in Clojure. Spark Streaming was developed by UC Berkeley,

whereas Storm was developed by BackType and Twitter (renamed X). Spark Streaming supports Scala and Java (and also Python), while Storm provides a Java API and supports APIs for other languages.

- *Execution Environment*: Both Spark Streaming and Storm can run in their respective cluster frameworks, but Storm can run on Mesos, while Spark Streaming can run on both YARN and Mesos. Since Spark Streaming runs on the Spark framework, developers can write Spark Streaming programs similarly to how they write other batch processing code, or perform interactive queries in Spark, reducing the need to write separate stream batch and historical data processing programs.

10.5.3. *GraphX*

GraphX is a distributed processing framework in Spark designed for graph processing (such as network graphs, web graphs, and social networks) and parallel graph computation (such as PageRank and Collaborative Filtering). It can be viewed as a rewrite and optimization of GraphLab (C++ implementation) and Pregel (C++ implementation) on Spark (Scala implementation). Compared to other distributed graph computation frameworks, GraphX's major contribution is providing a stack-based data solution on top of Spark, enabling a complete, efficient pipeline for graph computation. Initially developed as a distributed graph computation project by the Berkeley AMPLAB, GraphX later became a core component of Spark.

The scale and significance of graph data have driven the development of many parallel graph systems, such as Giraph and GraphLab. These graph computation models, by restricting the types of computations they can describe and introducing new graph partitioning methods, can effectively perform complex graph algorithms with much higher efficiency compared to more general data-parallel systems. GraphX in Spark is used for parallel graph computation, which involves splitting the graph into many subgraphs and computing each subgraph separately. The computations can be iterated in stages, allowing for parallel processing of the graph.

Graph computation is widely applied in social networks, such as Facebook and X, to calculate user connections. When the scale of a graph becomes very large, distributed graph computation frameworks are necessary, and GraphX, being a distributed graph processing framework based on Spark, naturally supports this requirement. The goal of GraphX's distributed framework is to package various operations on large-scale graphs into simple interfaces, making complex issues like distributed storage and parallel computation transparent to the user. This allows developers to focus more on model design and usage related to graph computation without worrying about the underlying distributed details.

GraphX extends Spark RDD with a core abstraction called the Resilient Distributed Property Graph (a directed multigraph with attributes on both vertices and edges). The Property Graph has two views—Table and Graph—but only one physical storage, consisting of VertexRDD and EdgeRDD. Each view has its own unique operators, making operations more flexible and improving execution efficiency. The structure of GraphX is shown in Figure 10.17.

Operations performed on the Graph view are ultimately converted into RDD operations on the associated Table view. Thus, computations on a graph are equivalent to a series of RDD transformations. Consequently, a Graph also inherits the three key characteristics of RDDs: immutability, distribution, and fault tolerance. Logically, all transformations and operations on a graph result in a new graph. Physically, GraphX applies optimizations to reuse immutable vertices and edges, though these optimizations are transparent to users.

The two views share underlying physical data, which consists of VertexRDD and EdgeRDD. Vertices and edges are not stored as table collections of tuples but rather as partitioned data blocks with indexing structures within VertexPartition and EdgePartition to accelerate traversal speeds across different views. This immutable indexing structure is shared during RDD transformations, reducing computation and storage overhead.

Table View   GraphX Unified Representation   Graph View

Fig. 10.17. Structure of GraphX.

Property Graph

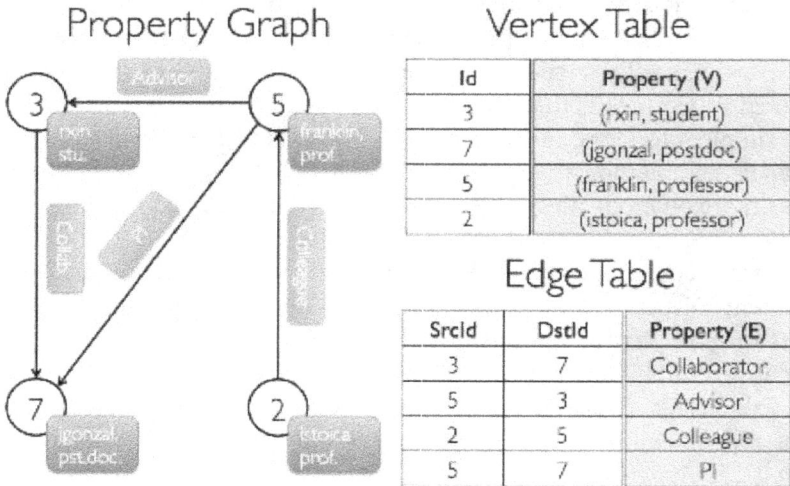

Vertex Table

Id	Property (V)
3	(rxin, student)
7	(jgonzal, postdoc)
5	(franklin, professor)
2	(istoica, professor)

Edge Table

SrcId	DstId	Property (E)
3	7	Collaborator
5	3	Advisor
2	5	Colleague
5	7	PI

Fig. 10.18.   Storage structure of GraphX.

The storage structure of GraphX is illustrated in Figure 10.18. In the Property Graph, there are two types of elements with attributes: vertices and edges. For example, vertices are labeled 3, 7, 5, and 2, and edges connect vertices, with arrows indicating relationships. By extracting vertices and edges, along with their attributes, the Property Graph can be transformed into a Vertex Table and an Edge Table. Vertex Table stores vertex data by converting vertices and their attributes into records. For instance, the Vertex Table includes records such as Id (representing vertices 3, 7, 5, and 2) and Property (V), which denotes the attributes associated with each vertex, like vertex 3 having attributes (rxin, student). Edge Table stores edge data by converting edges and their attributes into records. Each edge has associated SrcId and DstId representing the two vertices it connects. The Property (E) field denotes edge attributes, such as an edge connecting vertices 3 and 7 with an attribute of Collaborator, making Property (E) equal to Collaborator.

The distributed storage of graphs uses a vertex-cut model and employs the "partitionBy" method, allowing users to specify different partitioning strategies. These strategies assign edges to various "EdgePartitions" and distribute vertex masters to "VertexPartitions", with each "EdgePartition" caching local ghost copies of associated vertices. The choice of partitioning strategy affects the number of ghost copies that need to be cached and

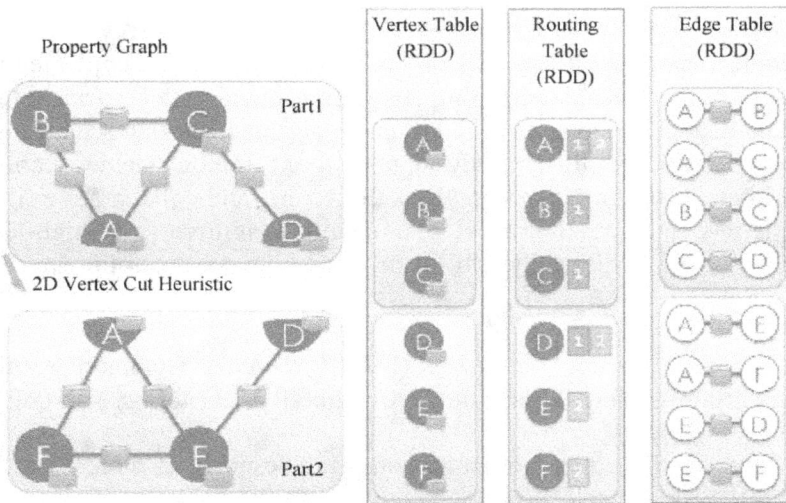

Fig. 10.19. Vertex-cut method.

the balance of edges allocated to each "EdgePartition". Selecting the optimal strategy depends on the structural characteristics of the graph. The vertex-cut approach stores graph data using three RDDs: "VertexTable", "RoutingTable", and "EdgeTable", as illustrated in Figure 10.19. A two-dimensional partitioning method based on vertex-cutting divides the graph into two partitions, "Part1" and "Part2", based on vertices A and D. Consequently, the vertex RDD ("Vertex Table RDD"), edge RDD ("Edge Table RDD"), and routing RDD ("Routing Table RDD") each contain two partitions. Vertex RDD partition 1 holds data for vertices A, B, and C, while vertex RDD partition 2 holds data for vertices D, E, and F. The routing RDD stores relationships between vertices and partitions; for instance, vertex A appears in both partition 1 and partition 2, while vertex B appears only in partition 1. Edge RDD partition 1 stores the edge data for "Part1", and edge RDD partition 2 stores the edge data for "Part2".

GraphX includes a set of graph algorithms designed to simplify analytical tasks. These algorithms are contained within the "org.apache.spark. graphx.lib" package and include well-known algorithms such as PageRank. The PageRank algorithm is used to determine the relative importance of an object within a graph dataset, measuring the significance of each vertex in the graph.

10.5.4. *Mllib*

Machine Learning Library (MLlib) is Spark's library for implementing commonly used machine learning algorithms, along with associated testing and data generators, designed to make machine learning scalable and more accessible. MLlib currently supports four common machine learning algorithms: classification, regression, clustering, and collaborative filtering. It also includes low-level optimization primitives and high-level pipeline APIs. Specifically, MLlib primarily covers the following five areas:

(1) *Machine Learning Algorithms* (*ML Algorithms*): Common learning algorithms, such as classification, regression, clustering, and collaborative filtering.
(2) *Featurization*: Feature extraction, transformation, dimensionality reduction, and selection.
(3) *Pipelines*: Tools for constructing, evaluating, and optimizing machine learning pipelines.
(4) *Persistence*: Saving and loading algorithms, models, and pipelines.
(5) *Utilities*: Tools for linear algebra, statistics, data processing, and more.

Starting from version 1.2, the Spark machine learning library is divided into two packages: "spark.mllib" and "spark.ml".

(1) spark.mllib contains the original RDD-based algorithm API. It has been around since before version 1.0, and the provided algorithms are implemented based on the original RDD.
(2) spark.ml offers a high-level API based on DataFrames, which can be used to build machine learning workflows (Pipeline). The ML Pipeline addresses some of the limitations of the original MLlib library, providing users with a DataFrame-based, workflow-oriented machine learning API suite.

Among all distributed architecture-based open-source machine learning libraries, MLlib can be considered one of the most computationally efficient. The main machine learning algorithms supported by MLlib are shown in Figure 10.20.

MLlib provides the following primary data types: local vector, labeled point, local matrix, and distributed matrix. It supports locally stored

	Discrete Data	Continuous Data
Supervised Learning	Classification, LogisticRegression(with Elastic-Net), SVM, DecisionTree, RandomForest, GBT, NaiveBayes, MultilayerPerceptron, OneVsRest	Regression, LinearRegression(with Elastic-Net), DecisionTree, RandomFores, GBT, AFTSurvivalRegression, IsotonicRegression
Unsupervised Learning	Clustering, KMeans, GaussianMixture, LDA, PowerIterationClustering, BisectingKMeans	Dimensionality Reduction, matrix factorization, PCA, SVD, ALS, WLS

Fig. 10.20.   Main machine learning algorithms supported by MLlib.

vectors and matrices in single-machine mode as well as distributed matrices based on one or more RDDs. Local vectors and matrices are offered as common interfaces in MLlib, providing simple data models, with linear algebra operations powered by the Breeze and jblas libraries (which are Spark's linear algebra libraries). The labeled point type is used to represent a training sample in supervised learning.

Exercises

(1) Please state the five main characteristics of RDDs.
(2) Briefly describe Spark's operation modes.
(3) What components are included in the Spark ecosystem?

Chapter 11

Cloud Computing Simulation

The first few chapters of this book have explained the knowledge related to virtualization-based cloud computing technology, cluster-based cloud computing technology, servers, and data centers. Based on these technologies, there are currently many system-level, algorithm-level, and application-level research approaches. Most of these developments and studies require simulation platforms. For example, technology developers conduct research on resource scheduling, load balancing, and cluster topology of large-scale clusters. If experiments are conducted on physical machines, it will inevitably consume a large number of hardware resources such as servers and network devices. The preparation of experimental environments, data collection, and debugging of experimental schemes are highly inconvenient, and the related costs are also quite high. In this case, using a simulation system is a good solution. For construction and operation personnel at data centers, energy consumption estimation and economic analysis of data centers are very important. It is necessary to make estimates before project construction. It is impossible to conduct calculations on actual platforms, and research needs to be conducted on simulation experiment platforms first.

This chapter focuses on CloudSim, a cloud computing simulation software. Through the study and use of this simulation software, readers can quickly master the relevant knowledge of cloud computing simulation.

11.1. Cloud Computing Simulation System: CloudSim

Among many cloud computing simulation systems, CloudSim is one of the most famous and widely used systems. The following is an introduction to CloudSim as an example of a cloud computing simulation system.

11.1.1. *CloudSim basics*

(1) *CloudSim Introduction*: CloudSim is a generic, scalable cloud computing simulation framework and a cloud computing simulation toolset developed by the Cloud Computing and Distributed Systems Laboratory at the University of Melbourne, Australia, providing core classes for describing data centers, virtual machines, applications, users, compute resources, and management policies.

The simulation of massive cluster resources has always been a focus of research in the computer field. Based on the CloudSim cloud computing simulation system, it is easy to not only build a controllable cloud computing environment and then model and test the resource scheduling and load balancing strategy of the system but also model and test cloud applications. R&D personnel can make targeted adjustments to performance bottlenecks based on the evaluation results. At the same time, CloudSim can establish pricing and energy consumption models for cloud computing systems, helping service providers develop more reasonable pricing strategies and energy-saving mechanisms.

Users can use the components provided by CloudSim to program and construct their own application scenarios, and they can also extend or write their own classes for simulation, which makes it very flexible to use. This is different from simulation systems for specific usage scenarios, which can be used without programming by filling in the parameters, but it is not possible to construct the usage scenarios in a flexible way.

CloudSim is developed in Java. Users only need to master the usage of Java and knowledge about cloud computing to build a cloud computing model for simulation. Of course, the simulation platform is a simulator and cannot run applications on a real cloud computing platform.

CloudSim performs resource allocation at both physical host and virtual machine levels. All virtual machines built in the physical host share physical resources, and the VmScheduler in CloudSim is responsible for resource allocation; the tasks simulated in CloudSim are called Cloudlets, and a large number of Cloudlets in the cluster require resources. The VM resource scheduler in CloudSim, called CloudletScheduler, is responsible for resource allocation.

(2) *Why use CloudSim*: For technology developers, research areas include resource scheduling, load balancing, cluster platform, and cluster topology. If such studies are carried out on physical machines, then a number of large-scale clusters, servers, and network equipment resources are required. Besides, the preparation of experimental environments, collection of experimental data, and debugging of experimental solutions are inconvenient and costly, so they need to carry out experiments on the simulation experiment platform first.

Testing of cloud application services can also be tricky in two main ways, as follows:

(1) Application service providers will undoubtedly incur additional cost overheads by testing applications after deploying them directly to the cloud platform. Once the application is connected to the cloud platform, it has to be paid for accordingly, which incurs additional costs without any economic benefits to the application and is not cost-effective for the SaaS provider.

(2) The actual running cloud platform environment (IaaS, PaaS, etc.) is uncontrollable, and the whole Internet environment is sometimes congested and sometimes idle, leading to irregular and non-reproducible use of cloud platform resources, which is not conducive to repeated testing of applications.

(3) *CloudSim Characteristics*:

- Capable of modeling and simulating large-scale cloud computing infrastructures, such as data centers and physical hosts, on a single PC.
- Support for modeling and simulation of user tasks and service agents.
- Support for modeling network environments in cloud computing environments.
- Effective use of virtualization engines to help create, manage, and destroy multiple virtual nodes on data center nodes.

- Flexibility to switch between time-share- and space-share-based virtualization policies.
- Support for modeling and simulation of energy consumption behavior in cloud data centers.
- Enabling easier establishment of a pricing strategy for cloud platform resources, including storage and bandwidth prices.
- Capable of mimicking transparent transactions between multiple cloud manufacturers, including task migration, storage migration, and price negotiation.

11.1.2. *CloudSim architecture*

The multilayer architecture of CloudSim is shown in Figure 11.1.

(1) *User Code Layer*: The user code layer is part of the upper layer of the system and contains simulation descriptions and scheduling policies, where the user defines the cloud computing scenarios and user

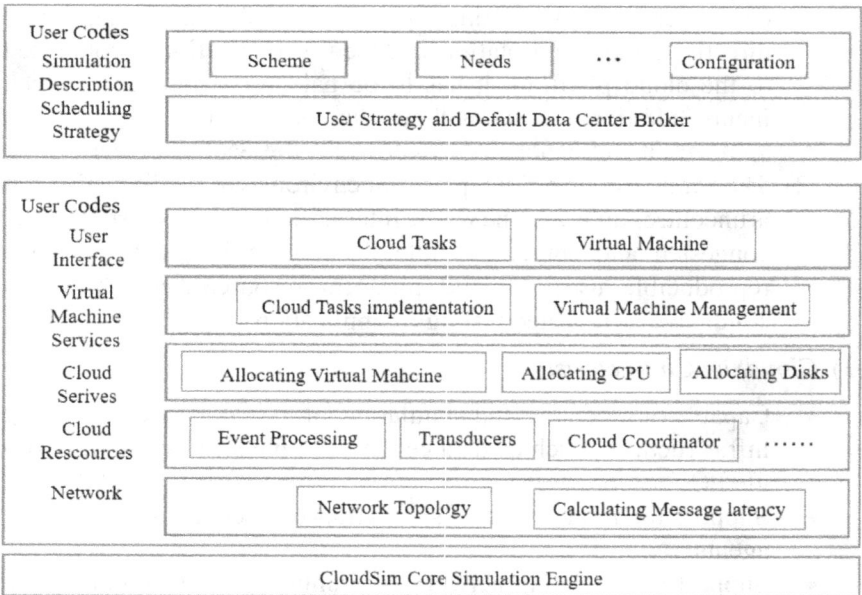

Fig. 11.1. CloudSim multi-layer architecture.

requirements and carries out the application configuration. At the same time, cloud application developers can generate workflow requests to test the cloud computing scenarios based on the user's configuration.

- *Simulation Description*: For cloud service users, it is necessary to test the performance of their applications on a specific cloud platform or to determine how much cloud resources their applications will require. Users only need to create a virtual cloud platform similar to the specific cloud platform and then create corresponding cloud tasks (defined as Cloudlets in CloudSim) according to the application's requirements (such as bandwidth and memory). Subsequently, they can run these cloud tasks on the virtual cloud platform and ultimately obtain the test results. For example, a user of the Amazon cloud platform who wants to deploy a network drive application and estimate the required service rental can use CloudSim for simulation. First, the user uses CloudSim to establish a virtual Amazon cloud platform; then, he creates a certain number of virtual machine resources on it, corresponding to a specific cloud service performance. Finally, the cloud service is generated according to the user's expectation (such as the required size of the hard disk, bandwidth, and memory), run it on the previously established virtual cloud service, and obtain the test results.

- *Scheduling Strategy*: From the perspective of cloud service providers, they may want to test whether the cloud platform's task scheduling strategy is reasonable or evaluate a new task scheduling strategy proposed by the service provider before it is implemented. The focus of testing, in this case, differs from that in CloudSim. The testing steps require the implementation of a custom task scheduling strategy, primarily by modifying the Data Center Broker. For instance, if an Amazon user finds that the current task scheduling strategy is not performing optimally, he can design and implement a new scheduling strategy and simulate it using CloudSim. First, he rewrites the task scheduling strategy code of the Data Center Broker. Then, he creates the cloud platform and cloud tasks, runs the simulation, and ultimately obtains the test results.

(2) *CloudSim Simulation Layer*: The primary function of the CloudSim simulation layer is to model and simulate virtual machines, memory,

storage, bandwidth, and other resources in a virtualized data center environment. Tasks such as partitioning physical machines into virtual machines, application management, and cluster system status monitoring are handled by the CloudSim simulation layer. Users can write their own policies within the CloudSim simulation layer to study and evaluate the virtual host allocation strategies in a virtualized data center. This allows them to assess the performance of the data center under different allocation strategies. Cloud application developers can also use the CloudSim simulation layer to test the performance of various cloud applications.

The basic building blocks of an actual cloud computing environment are data centers. A data center contains a large number of physical hosts, and in a cloud environment, these physical hosts can be shared by multiple virtual machines. CloudSim defines a set of resource-sharing policy interfaces (UtilizationModel) to describe how shared resources are utilized. In CloudSim, a host can be shared by multiple virtual machines. The primary resource-sharing policies include space-sharing and time-sharing strategies.

The space-sharing strategy refers to allocating computing resources exclusively to a single virtual machine or computing task for a certain period of time. In contrast, the time-sharing strategy allows computing resources to be shared among multiple virtual machines or computing tasks during the same time period. For example, consider a situation where a host owns two CPUs. CloudSim deploys two virtual machines, VM1 and VM2, on this host. Each virtual machine has four tasks: VM1 has tasks t1, t2, t3, and t4, while VM2 has tasks t5, t6, t7, and t8, as illustrated in Figure 11.2.

Figure 11.2(a) shows the timeline of computing tasks where both the host and virtual machine layers adopt the space-sharing strategy. In this scenario, VM1 initially exclusively occupies the two CPUs, completing its tasks before handing over control to VM2. During this time, tasks t1 and t2 each exclusively use CPU1 and CPU2, respectively. Once these tasks are completed, t3 and t4 are executed.

Figure 11.2(b) illustrates the scenario where the host layer uses the space-sharing strategy while the virtual machine layer uses the time-sharing strategy. Figure 11.2(c) depicts the scenario where the host layer uses the time-sharing strategy while the virtual machine layer uses the

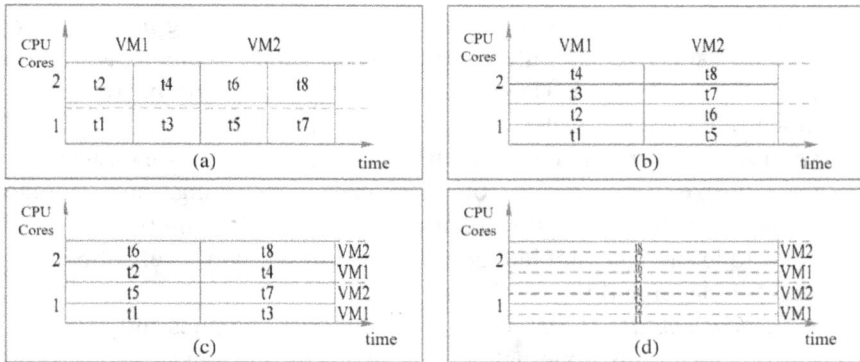

Fig. 11.2. Task execution under different resource sharing strategies.

space-sharing strategy. Figure 11.2(d) shows the scenario where both the host and virtual machine layers adopt the time-sharing strategy.

11.2. CloudSim Model Usage Scenarios

CloudSim can simulate many aspects of cloud data centers, such as networks, power, and the operation of virtual machines. The models in CloudSim mainly fall into two categories: the cloud data center energy consumption model and the cloud data center economic model.

(1) *Cloud Data Center Energy Consumption Model*: Cloud data centers consist of a large number of interconnected hosts, storage devices, and network equipment. Maintaining the operation of such a vast system requires a considerable amount of electricity. CloudSim provides simulation of power control strategies, enabling users to design power schemes that align with the characteristics of local data centers, thereby saving costs and improving the overall efficiency of the system.

In CloudSim, an abstract class called PowerModel has been implemented to model power strategies. Users can inherit from this abstract class to write their own cloud data center power supply schemes and conduct simulation experiments on CloudSim to validate the overall effectiveness of the power supply scheme.

(2) *Cloud Data Center Economic Model*: Cloud computing provides services to users over the Internet, offering dynamic, scalable, and often virtualized resources. Users can access cloud computing resources, much like they use water and electricity, simply by paying the cloud service provider for the computing, storage, and network resources they utilize. Pricing for computational, networking, and storage resources is crucial for the operation of cloud data centers.

CloudSim can simulate pricing strategies at two levels: the infrastructure layer and the service layer.

- *Infrastructure layer*: This layer primarily includes the prices of memory units, external storage, unit costs of data transmission, and computing resources.
- *Service layer*: This layer mainly pertains to the prices of resources used by application services. If the user only utilizes the infrastructure of the cloud data center without deploying any applications on it, or merely creates a few virtual machines without running any tasks on them, then there is no need to pay for the service layer.

CloudSim's Datacenter class includes several parameters related to pricing, such as the usage prices for CPU, network, memory, and storage. These parameters make it convenient for users to model the pricing strategies of a cloud data center.

11.3. CloudSim Application and Practice

CloudSim is an open-source software written in Java, requiring a properly set up Java runtime environment to run. Here, the author uses CloudSim integrated with Eclipse to conduct simulation experiments and development work in cloud computing, with the operating system being Windows 7. This section covers the following aspects: setting up and testing the CloudSim environment, a data center simulation instance, and a network simulation instance. The data center simulation instance is used to simulate the operation of a data center, while the network simulation instance is used to simulate the operation of cloud transactions under network latency conditions.

11.3.1. *Preparation environment*

(1) *Download CloudSim*: Visit http://code.google.com/p/cloudsim/ downloadstodownload"cloudsim-3.0.3.zip" and extract it.

(2) *Preparing for Eclipse Development Environment*:
- Based on the CPU of the user's machine, download and install the appropriate version of Eclipse.
- Click on "File" → "New" → "Java Project" to create a new Java project. Name the project "CloudSim", as shown in Figure 11.3.
- Since this project uses classes from the math library, it is necessary to include the commons-math3-3.2.jar library. After downloading commons-math3-3.2.jar, select the newly created project

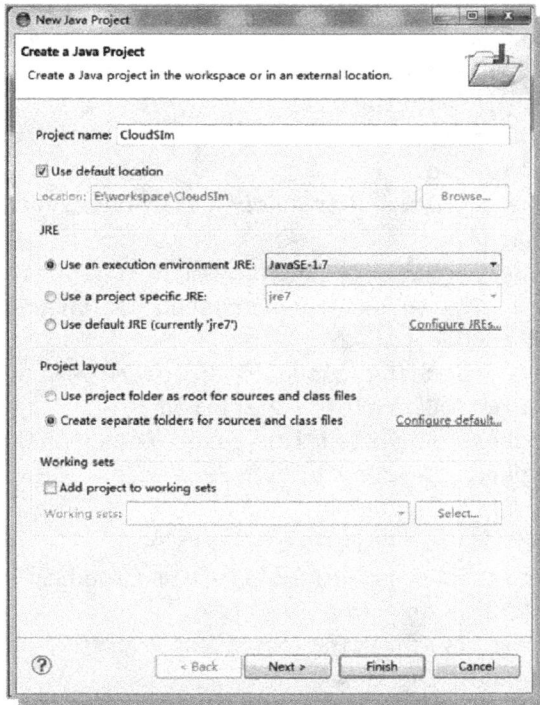

Fig. 11.3. Preparing the eclipse development environment.

"CloudSim", right-click on it, and choose "Build Path" → "Add External Archives" to import the library.

(3) *Running Testing Program*: CloudSim provides some example programs to help beginners quickly understand CloudSim. These example programs are stored in the /cloudsim-3.0.3/examples/org/cloudbus/cloudsim/examples directory after extracting the CloudSim folder. Open this directory and copy the six example programs from CloudSimExample1.java to CloudSimExample6.java and into the project. Here, the author considers the example program CloudSimExample6.java. Press Ctrl+F1 to run the example program. After running, if the reader sees the following output, it indicates that the CloudSim environment has been successfully set up and can be used normally:

```
Starting CloudSimExample1.
Initialising…
StartingCloudSimversion3
Datacenter0isstarting…
Brokerisstarting…
Entities started.
0.0: Broker: Cloud Resource List received with 1
    resource(s)
0.0: Broker: Trying to Create VM #0 Datacenter_0
0.1: Broker: VM #0 has been created in Datacenter #2,
    Host #0
0.1: Broker: Sending cloudlet 0 to VM #0
400.1: Broker: Cloudlet 0 received
400.1: Broker: All Cloudlets executed. Finishing…
400.1: Broker: Destroying VM #0
Broker is shutting down…
Simulation: no more future events
CloudIntormationsezvice: Notify a11 CloudSim entities
    for shutting down.
Datacenter_0 is shutting down…
Broker is shutting down…
Simulation completed.
Simulation completed.
```

```
==========OUTPUT==========
Cloudlet ID STATUS Data center ID VM ID Time Start
   Time Finish Time
  0 SUCCESS 2 0 400 0.1 400.1
CloudSimExample1 finished!
```

11.3.2. *Data center simulation instance*

In this section, the author uses CloudSim to simulate a data center. The data center consists of a cluster of two dual-core physical machines, with each physical machine hosting four virtual machines. This means that two virtual machines share one CPU core. In total, there are eight virtual machines in the cluster, each with different computing capacities (MIPS). The data center needs to handle an external workload of 16 tasks.

(1) *Creating Virtual Machine*: In CloudSim, the performance of virtual machines can be defined using parameters such as image size, memory size, CPU processing power, and network bandwidth. The code to create virtual machines is as follows:

```
/** The method of creating virtual machine */
private static List<Vm> createVM(int userId){
    // create a chained table to store the created VMs
    LinkedList<Vm> list = new linkedList<Vm>();
    /* parameters of virtual machine */
    long size = 10000;  // size of image(MB)
    int ram = 512;      //memory of virtual machine(MB)
    long bw = 1000;     // band width(KBPS)
    int pesNumber = 1;  // the number of the virtual
                        machine cores
    String vmm = "Xen"; // the type of the virtual
                        machine
    Vm[] vm = new Vm[mips.length];
    for(int i = 0; i< mips.length; i++){
        // create virtual machine based on time-sharing
          policies
```

```
        vm[i] = new Vm(i, userId, mips[i], pesNumber,
            ram, bw, size, vmm,
            new CloudletSchedulerTimeShared());
        list.add(vm[i]);
    }
    return list;
}
```

(2) *Creating Cloud Transaction Tasks*: Next, create the cloud transaction task and define the execution duration, occupied space size, output file size, and number of CPU cores used for the task, as follows:

```
/** The method of creating cloud transaction tasks */
private   static   List<Cloudlet>   createCloudlet(int
    userId, long cloudlets[]){
    // create a chained table to store cloud transactions
    LinkedList<Cloudlet> list = new
        linkedList<Cloudlet>();
    /* parameters of cloud transaction tasks */
    long fileSize = 300;      // size of the files(MB)
    long outputSize = 300;    // size  of  the  output
                              files(MB)
    int pesNumber = 1;        // the  number  of  the
                              virtual machine cores

    /* strategy of sharing resources */
    UtilizationModel   utilizationModel   =   new
        UtilizationModelFull();
    Cloudlet[]  cloudlet  =  new  Cloudlet[Cloudlets.
        length];
    for(int i = 0; i<10; i++){
        cloudlet[i]   =   new   Cloudlet(i,  length[i],
            pesNumber, fileSize,
            outputSize, utilizationModel, utilizationModel,
                utilizationModel);
        // set up who the cloud transactions belong to
        cloudlet[i].setUserId(userId);
        list.add(cloudlet[i]);
    }
    return list;
}
```

(3) *Running the Main Program*: The main program is the focus of CloudSim simulation, and the main steps to run the CloudSim simulation main program are divided into six steps: initialize the CloudSim package, create the data center, create the data center agent, create the virtual machine and cloud transaction, start the simulation, and print the simulation results, as shown in the following:

```
/* The instance of creating a main function*/
public static void main(String[] args){
  Log.printLine("Starting CloudSimExampleA...");
  try{
    /* Step 1: Initialise the CloudSim package
       before all entities are created */
    int num_user = 1;           //the numbers of
                                  the user
    Calendar calendar =Calendar.getInstance();
    boolean trace_flag = false; //whether to follow
                                  the evens
    /* initialize CloudSim library*/
    CloudSim.init(num_user, calendar, trace_flag);
    /* Step 2: Create a data center*/
    Datacenter datacenter0 =createDatacenter("Datac
      enter_0");
    /* Step 3: Create data centre (user) agents */
    DatacenterBroker broker = createBroker();
    int brokerId = broker.getId();
    /* Step 4: Create VMs, cloud transactions and
       pass them to the data centre */
    int mips[] = {278, 289, 132, 209, 286, 333, 212,
      423 };                    // CPU performance
                                  of the virtual
                                  machine (MIPS)
    /* number of instructions required */
    long cloudlets[] = new long[] { 19365, 49809,
      30218, 44157, 16754, 18336, 20045, 31493,
      30727, 31017, 59008, 32000 ,46790, 77779,
      93467, 67853 };
    vmlist = createVM(brokerId);
    cloudletList = =createCloudlet(brokerId);
```

```
broker.submitVmList(vmlist);
broker.submitCloudletList(cloudletList);
/* Step 5: Start simulation*/
CloudSim.startSimulation();
/* Step 6: End of simulation and print the
    simulation result*/
List<Cloudlet>      newList    =      broker.
    getCloudletReceivedList();
CloudSim.stopSimulation();
printCloudletCost(newList);
Log.printLine("CloudSimExample6 finished!");
} catch (Exception e){
    e.printStackTrace();
    Log.printLine(
    "The simulation has been terminated due to an
        unexpected error.");
    }
}
```

The running results are shown in the following:

```
Starting CloudSimExampleA...
Initialising...
Starting CloudSim version 3.0
Datacenter_0 is starting...
Broker is starting...
Entities started.
0.0: Broker: Cloud Resource List received with 1
    resource(s)
0.0: Broker: Trying to Create VM #0 in Datacenter_0
0.0: Broker: Trying to Create VM #1 in Datacenter_0
0.0: Broker: Trying to Create VM #2 in Datacenter_0
0.0: Broker: Trying to Create VM #3 in Datacenter_0
0.0: Broker: Trying to Create VM #4 in Datacenter_0
0.0: Broker: Trying to Create VM #5 in Datacenter_0
0.0: Broker: Trying to Create VM #6 in Datacenter_0
0.0: Broker: Trying to Create VM #7 in Datacenter_0
0.1: Broker: VM #0 has been created in Datacenter #2,
    Host #0
```

0.1: Broker: VM #1 has been created in Datacenter #2, Host #1
0.1: Broker: VM #2 has been created in Datacenter #2, Host #0
0.1: Broker: VM #3 has been created in Datacenter #2, Host #1
0.1: Broker: VM #4 has been created in Datacenter #2, Host #0
0.1: Broker: VM #5 has been created in Datacenter #2, Host #1
0.1: Broker: VM #6 has been created in Datacenter #2, Host #0
0.1: Broker: VM #7 has been created in Datacenter #2, Host #1
0.1: Broker: Sending cloudlet 0 to VM #0
0.1: Broker: Sending cloudlet 1 to VM #1
0.1: Broker: Sending cloudlet 2 to VM #2
0.1: Broker: Sending cloudlet 3 to VM #3
0.1: Broker: Sending cloudlet 4 to VM #4
0.1: Broker: Sending cloudlet 5 to VM #5
0.1: Broker: Sending cloudlet 6 to VM #6
0.1: Broker: Sending cloudlet 7 to VM #7
0.1: Broker: Sending cloudlet 8 to VM #0
0.1: Broker: Sending cloudlet 9 to VM #1
0.1: Broker: Sending cloudlet 10 to VM #2
0.1: Broker: Sending cloudlet 11 to VM #3
0.1: Broker: Sending cloudlet 12 to VM #4
0.1: Broker: Sending cloudlet 13 to VM #5
0.1: Broker: Sending cloudlet 14 to VM #6
0.1: Broker: Sending cloudlet 15 to VM #7
110.22073584375741: Broker: Cloudlet 5 received
117.25570087872245: Broker: Cloudlet 4 received
139.41397426001743: Broker: Cloudlet 0 received
149.00262674228694: Broker: Cloudlet 7 received
180.28320228185527: Broker: Cloudlet 8 received
189.1982966214779: Broker: Cloudlet 6 received
214.74775417625537: Broker: Cloudlet 9 received
222.27643507571565: Broker: Cloudlet 12 received

```
234.95804857081453: Broker: Cloudlet 15 received
279.77119735974185: Broker: Cloudlet 1 received
288.7291553176998: Broker: Cloudlet 13 received
306.31767206411126: Broker: Cloudlet 11 received
364.48256020450873: Broker: Cloudlet 3 received
457.93710565905417: Broker: Cloudlet 2 received
535.5267283005636: Broker: Cloudlet 14 received
676.0418798157151: Broker: Cloudlet 10 received
676.0418798157151: Broker: All Cloudlets executed.
    Finishing...
676.0418798157151: Broker: Destroying VM #0
676.0418798157151: Broker: Destroying VM #1
676.0418798157151: Broker: Destroying VM #2
676.0418798157151: Broker: Destroying VM #3
676.0418798157151: Broker: Destroying VM #4
676.0418798157151: Broker: Destroying VM #5
676.0418798157151: Broker: Destroying VM #6
676.0418798157151: Broker: Destroying VM #7
Broker is shutting down...
Simulation: No more future events
CloudInformationService: Notify all CloudSim entities
    for shutting down.
Datacenter_0 is shutting down...
Broker is shutting down...
Simulation completed.
Simulation completed.
```

```
========== OUTPUT ==========
```

Cloudlet ID	STATUS	Data center ID	VM ID	Time	Start Time	Finish Time
5	SUCCESS	2	5	110.12	0.1	110.22
4	SUCCESS	2	4	117.16	0.1	117.26
0	SUCCESS	2	0	139.31	0.1	139.41
7	SUCCESS	2	7	148.9	0.1	149
8	SUCCESS	2	0	180.18	0.1	180.28
6	SUCCESS	2	6	189.1	0.1	189.2
9	SUCCESS	2	1	214.65	0.1	214.75
12	SUCCESS	2	4	222.18	0.1	222.28

```
15          SUCCESS  2      7     234.86    0.1 234.96
1           SUCCESS  2      1     279.67    0.1 279.77
13          SUCCESS  2      5     288.63    0.1 288.73
11          SUCCESS  2      3     306.22    0.1 306.32
3           SUCCESS  2      3     364.38    0.1 364.48
2           SUCCESS  2      2     457.84    0.1 457.94
14          SUCCESS  2      6     535.43    0.1 535.53
10          SUCCESS  2      2     675.94    0.1 676.04
CloudSimExampleA finished!
```

In order to understand the results of its operation, the reader can start by focusing on the fourth step of the main program code, noting in particular the following two parts of the code:

(1) *CPU Performance*:

```
int mips[] = new int[] {278, 289, 132, 209, 286, 333,
   212, 423 };   // CPU performance of the virtual
   machine
```

(2) *The Number of Instructions Executed by the Task*:

```
long cloudlets[] = new long[] { 19365, 49809, 30218,
   44157, 16754, 18336,
20045, 31493, 30727, 31017, 59008, 32000, 46790,
   77779, 93467, 67853 };
```

Examining the results, note the completion of transactions 5 and 2. Transaction 5 was allocated to VM5, and transaction 2 was allocated to VM2. The CPU performance values for VM2 and VM5 are 289 and 286, respectively, indicating that their CPU execution efficiency is very similar and can be considered the same for this analysis. The number of instructions for transactions 2 and 5 are 16,754 and 49,809, respectively, and their execution times are 110.22 and 676.04 s, respectively. From this result, we can infer that, given the same CPU computation capability, the number of instructions executed determines the transaction processing time. Similarly, let's consider another example: transactions 9 and 3. The number of instructions for transactions 9 and 3 are 30,727 and 30,218, respectively, which are very close and can be regarded as the same. Transactions 9 and 3 are allocated to VM1 and VM3, respectively. The CPU performance values for VM1 and VM3 are 278 and 132, respectively. From the results, we see that transaction 9 took 214.65 s to complete, while transaction 3 took 364.38 s. Therefore, we can conclude that, given the same transaction length, the CPU execution speed (i.e., CPU computation capability) determines the transaction processing time.

At this point, the author has completed the simulation experiment of a small data center consisting of two physical machines. From the simulation results above, it is clear that CPU computation capability and transaction processing time directly impact the duration of task execution. With the same transactions, stronger CPU computation capabilities lead to shorter execution times. Conversely, with the same CPU computation capability, shorter transactions result in less time consumed. Therefore, it is crucial to efficiently utilize virtual machine computational resources and to allocate tasks reasonably.

11.3.3. *Network simulation example*

This section introduces how to use CloudSim for network simulation and to run cloud transactions (in this context, cloud transactions refer to tasks or loads), simulating network traffic latency. In network simulation, it is necessary to create hosts and run data centers on them, then run cloud transactions on the network topology. The operations of creating virtual machines, creating cloud transactions, and defining data centers are the same as in the previous examples. For simplicity, this example uses only one data center and one network topology.

The network topology in CloudSim relies on BRITE files, which can contain many entity nodes. This allows users to scale the simulation without changing the topology file. Each CloudSim entity must be mapped to one (and only one) node to ensure that the network simulation works correctly, and each BRITE node can only be mapped to one entity at a time. The following is an example of the content of a BRITE file:

```
Topology: ( 5 Nodes, 8 Edges )
Model (1 - RTWaxman): 5 5 5 1 2 0.15000000596046448
   0.20000000298023224 1 1 10.0 1024.0
Nodes: ( 5 )
0 1 3 3 3 -1    RT_NODE
1 0 3 3 3 -1    RT_NODE
2 4 3 3 3 -1    RT_NODE
```

```
3  3  1  3  3  -1     RT_NODE
4  3  3  4  4  -1     RT_NODE
Edges:  ( 8 )
0  2  0  3.0                        1.1   10.0  -1  -1   E_RT  U
1  2  1  4.0                        2.1   10.0  -1  -1   E_RT  U
2  3  0  2.8284271247461903         3.9   10.0  -1  -1   E_RT  U
3  3  1  3.605551275463989          4.1   10.0  -1  -1   E_RT  U
4  4  3  2.0                        5.0   10.0  -1  -1   E_RT  U
5  4  2  1.0                        4.0   10.0  -1  -1   E_RT  U
6  0  4  2.0                        3.0   10.0  -1  -1   E_RT  U
7  1  4  3.0                        4.1   10.0  -1  -1   E_RT  U
```

The information affecting the simulation begins with "Nodes". Each row under Nodes represents a node in the topology, with each column respectively indicating the following: node ID, x-coordinate position, y-coordinate position, in-degree, out-degree, autonomous system ID, and type (routing/autonomous). Each row under Edges represents an edge in the topology, with each column respectively indicating the following: edge ID, start point, end point, Euclidean distance, delay, bandwidth, autonomous system start, autonomous system end, and type.

CloudSim uses the NetworkTopology class to implement network layer functionality. The NetworkTopology class can read BRITE files and generate a network topology from them. This information is used to simulate network traffic delay in CloudSim.

The methods for creating data centers and cloud transactions will not be elaborated upon here. The following is the main program for network simulation:

```java
public static void main(String[] args) {
    Log.printLine("Starting NetworkExampleA...");
    try {
        // Step 1: Initialize the CloudSim package
        before creating all entities.
        int num_user = 1;                // number of the
                                         users
        Calendar calendar = Calendar.getInstance();
        boolean trace_flag = false;      // whether  to
                                         follow the events
```

```
// Initialize CloudSim class
CloudSim.init(num_user, calendar, trace_flag);

// Step2: Create a data center
//Datacenters are the resource providers in
  CloudSim. We need at list one of them to run
  a
CloudSim simulation
Datacenter datacenter0 = createDatacenter("Data
  center_0");

// Step 3: Create data center (user) agents
DatacenterBroker broker = createBroker();
int brokerId = broker.getId();

// Step4: Create a VM
int mips[] = {278};
vmlist = createVM(brokerId,mips);
broker.submitVmList(vmlist);

// Step5: Create a cloud event event
long length[] = { 40000 };
cloudletList = createCloudlet(brokerId,length);
  // creating 40 cloudlets
broker.submitCloudletList(cloudletList);

// Step 6: Load network configuration parameters
NetworkTopology.buildNetworkTopology("topology.
  brite");
// map CloudSim entities to BRITE entities.
int briteNode=0;
NetworkTopology.mapNode(datacenter0.
  getId(),briteNode);
// map the broker to BRITE entity number 3
briteNode=3;
NetworkTopology.mapNode(broker.
  getId(),briteNode);

// Step 7: Start the simulation
CloudSim.startSimulation();
```

```
    // Step 8: End the simulation and print the
        results.
    List<Cloudlet> newList = broker.
        getCloudletReceivedList();

    CloudSim.stopSimulation();
    printCloudletList(newList);
    Log.printLine("NetworkExampleA finished!");
  }
  catch (Exception e) {
    e.printStackTrace();
    Log.printLine("The simulation has been terminated
        due to an unexpected error");
  }
}
```

The program's execution results are as follows:

```
Starting NetworkExampleA...
Initialising...
Topology file: topology.brite
Starting CloudSim version 3.0
Datacenter_0 is starting...
Broker is starting...
Entities started.
0.0: Broker: Cloud Resource List received with 1
    resource(s)
7.800000190734863: Broker: Trying to Create VM #0 in
    Datacenter_0
15.700000381469726: Broker: VM #0 has been created in
    Datacenter #2, Host #0
15.700000381469726: Broker: Sending cloudlet 0 to VM
    #0
167.3848926585355: Broker: Cloudlet 0 received
167.3848926585355: Broker: All Cloudlets executed.
    Finishing...
167.3848926585355: Broker: Destroying VM #0
```

```
Broker is shutting down...
Simulation: No more future events
CloudInformationService: Notify all CloudSim entities
   for shutting down.
Datacenter_0 is shutting down...
Broker is shutting down...
Simulation completed.
Simulation completed.

========== OUTPUT ==========
Cloudlet STATUS  Data        VM Time    Start Finish
ID                center ID  ID         Time  Time

  0      SUCCESS 2           0  143.88 19.6  163.48
NetworkExampleA finished!
```

According to the output results above (the content under "OUTPUT"), the start time of cloud transactions (Start Time) is 19.6, instead of the 0.1 as observed in the data center simulation instance (refer to the data center simulation instance results). This indicates that the start time is directly influenced, while the execution time remains unaffected, implying the presence of network latency during the simulation of network-operated transactions. In a non-network simulation environment, the start time of cloud transactions should be 0.1.

Exercises

(1) What is CloudSim?
(2) What are the model scenarios used by CloudSim?
(3) Briefly describe the main steps of CloudSim simulation.
(4) Use CloudSim to simulate the following data center: Simulate two data centers, each with 10 physical machines (five dual-core and five quad-core). There are a total of 100 virtual machines in the two data centers, each with different computing capabilities, ranging from 100 to 500. These two data centers need to handle a total of 1000 external load tasks, with load capacities ranging from 10,000 to 100,000.

Index